Sleeps *with* Dogs

Tales of a Pet Nanny
at the End of Her Leash

LINDSEY GRANT

SEAL PRESS

Sleeps with Dogs
Copyright © 2014 Lindsey Grant

Author's Note: I have tried to re-create the events, locales, and conversations included in this book from my memories of them. In order to maintain their anonymity, I have changed the names of individuals and the animals described here, as well as some identifying characteristics and details such as geographic descriptions, physical properties, occupations, and places of residence.

Published by
Seal Press
A Member of the Perseus Books Group
1700 Fourth Street
Berkeley, California

www.sealpress.com

Library of Congress Cataloging-in-Publication Data

Grant, Lindsey, 1982- author.
  Sleeps with dogs : tales of a pet nanny at the end of her leash / Lindsey Grant.
    pages cm
  ISBN 978-1-58005-547-5 (paperback)
  1. Pets—Anecdotes. 2. Pet owners—Anecdotes. 3. Grant, Lindsey, 1982- I. Title.
  SF416.G73 2014
  636.088'7—dc23
                              2014025354
10 9 8 7 6 5 4 3 2 1

Cover design by Raquel Van Nice
Interior design by Domini Dragoone
Printed in the United States of America
Distributed by Publishers Group West

For my parents, who encouraged every beginning,
and for P, who made "The End" possible

Dear Neighbor,

I am pleased to introduce myself as a new name in local animal care. While I am only recently a resident of California, I'm an old hand when it comes to working with (and doting upon) animals.

It's my pleasure to offer my services to you and your pet—canine, feline, reptilian, avian, or otherwise—with the guarantee that, while I am with your companion, my top priorities are always safety, consistent quality of care, and lots and lots of affection!

I look forward to meeting you and your (furry, scaly, or feathered) friend!

Lindsey Grant

CHAPTER ONE

Will Work for Pets

....

I was the go-to kid on the block to call when the neighbors were going out of town and needed their newspapers brought up, their mail retrieved, the plants watered, and their dogs walked or their cats fed while they were away. Before I was old enough to babysit, and even after, I regularly watched after the pets living in my family's intown Atlanta neighborhood.

For years, I pet-sat for a Lab and an aging cocker spaniel down the street, never quite getting used to the idea that I'd earned the crisp $50 bill the neighbors always paid me with when they returned. I'd have played with those dogs, pet them, fed them, and refreshed their water bowls, for free.

As the spaniel got older and older, she got more and more ornery, once biting me on the face when I leaned in too close to greet her. I was devastated, certain that I'd done something really wrong to prompt her to bite me like that. In my third-grade class, we'd had a hamster that bit me, and I never forgot the teacher's explanation that animals bite because they are afraid and are trying to protect themselves. I'd cried about that hamster bite, not because it hurt, but because I felt so bad for scaring the hamster.

After the spaniel attack, I left a note for her owners explaining what happened and suggesting that they shouldn't pay me because I had scared her and I was sorry. They of course paid me anyway; I heard them at the door talking to my parents and apologizing profusely for what had happened. The dog was getting old and senile, they'd said, and they'd been worried for a while about their own young kids playing with her for this reason.

I also took care of two big beautiful golden retrievers who, unlike our dog, Biscuit, lived inside the house and ate wet Alpo, which smelled much worse than Biscuit's dry pellets. And once, I took care of our family friends' budgies when they went on vacation. I couldn't understand the appeal of those loud, demanding birds—also biters—but I was happy to be asked.

Even when I started taking on babysitting gigs, I so much preferred the demands of the dogs and birds over the kids I cared for. Animals didn't complain about the way you prepared their SpaghettiOs, balk about bed or bath times, or question the necessity of tooth brushing. They didn't challenge your authority by throwing things at your head or tell you they hated you and wanted their mommy. Increasingly, I declined requests to babysit or nanny and just stuck with the animals.

While I grew up with traditional pets—seven or so hamsters, our beloved Biscuit, and Seal the cat—I had a snake in college. Monty was a ball python, about two feet long, and a wonderful companion. As pets go, snakes aren't the most expressive or the cuddliest. But to me, he was both. He never bit me and was content to hang out on the couch, wrapped around my arm or coiled under a throw pillow, while I studied or watched TV. He was exceedingly polite to guests and even acted like a gentleman in the company of the live rats I fed him. If he was hungry, he dispatched them quickly and cleanly. A hero's death. If not, he endured the rat's fearful bites nobly, never retaliating when his rat snack sunk its teeth into his perfect scaly length.

Perversely, I grew attached to the rats, too—the uneaten ones. When Monty declined his dinner, I'd keep the rat in a shoe box with a cardboard tube and an ear of corn or some cereal until my next shift at the strip mall pet store, where I'd worked since sophomore year. There, the rejected rat, spared its inevitable fate for one day more, would go back in the tank with his rodent brethren.

After college, my move from Georgia to California necessitated that I find Monty a new home. Taking him with me on the cross-country journey hadn't even occurred to me. This was 2004, and *Snakes on a Plane* hadn't yet been released. I didn't investigate at the time, certain that there weren't allowances for reptilian carry-ons. Nevertheless, as soon as I relinquished Monty to my manager at the pet store, along with his hand-painted cage of wood and chicken wire, heat lamp, water bowl, and bark arch for hiding beneath, I sorely wished I'd reconsidered.

Snakes don't exactly give you that heart-squeezing look that you might get from a beloved dog, or cat, or rabbit. He flicked his tongue, testing the stale air in the back of the store, no doubt sensing the proximity of many other snakes and dozens of tasty rodent treats. Surely, his tiny snake heart was not hurting like mine did during this final farewell. My former boss was more than happy to take my handsome two-foot-long friend off my hands, certain that she could sell him for a nice sum.

Once I'd surrendered my snake, I packed up the rest of my belongings for shipping West. Beyond the bedding and clothes, my boxes were filled with largely impractical and sentimental creature comforts—books, my parents' decades-old vinyl collection and turntable, framed photos. I was proud that I'd condensed twenty years of living in the Atlanta area into roughly six boxes of varying sizes, until I saw the amount I owed for shipping. Perhaps I could have reconsidered some of the books and all of the records.

My plan was to establish California residency in order to attend grad school. Creative writing seemed a likely fit for my interests and skills; far better than a master's in linguistics or a PhD in literature. I'd studied literature and film at the University of Georgia, a state school liberally attended by people I'd grown up with. Four years of college parties and rooming with kids I'd known since elementary school, in a town where alcohol poisoning was de rigueur and every place seemed overrun with pledges and their Greek sisters and brothers, was enough to instill in me a potent fear of stagnation and mediocrity, and the desire for a significant change. Moving to California instead of returning to Atlanta post-graduation seemed like a pretty good, if dramatic, way to avoid the former and achieve the latter.

Plans rarely pan out; I knew this. But that was mine, and I was sticking to it. My mom's best friend from childhood lived southeast of Berkeley with her husband and two young boys, and she had offered to help me get on my feet. She was an English professor and had been providing me with ample information about the excellent writing programs in the area. It was to her address that I sent my belongings, with her that I shared my flight info. I had never before purchased a one-way ticket. I was leaving my Atlanta home to go West, to write.

Annie had set up a pet-sitting gig for me with her neighbors down the street. They were off to visit family for two weeks and needed a house sitter to take care of their dogs. Two weeks sounded like the perfect amount of time to get my bearings, apply for work, and figure out the year that yawned before me.

The neighbors offered me $30 a day; more money than I could wrap my brain around for hanging out with two dogs, feeding and petting and walking them, and sleeping over at their gorgeous Craftsman home.

At the pet store, I'd been paid minimum wage to show up at six o'clock in the morning and clean each puppy cubby, flushing pounds and pounds of dog shit down an industrial disposal. Once all of the cages had been sanitized, first with bleach and then with a highly concentrated pink solution to protect against parvo, the dogs could be restored to their now-habitable display cases, and we could open the store to the clamoring hordes. During the day, it was my job to keep the trays clean of any leavings and leap into action if any of the dogs started rolling in—or worse, eating—their turds. My colleagues working the floor—"Pet Counselors," their nametags declared—would rap on the kennel door and call "retriever" or "Dalmatian," and I'd scurry down the line to deal with it. Beyond that, I fed, medicated, groomed, laundered, and generally kept things clean and running smoothly behind the scenes.

Disgusting a job as it could prove at times, I loved those animals. The grand equalizer was the time I spent walking or brushing or snuggling them. I worried when one of my favorites was quarantined with a cold, and I missed each and every dog or cat that went home with a customer. For those few years, I was the den mother and they were my cubs.

Getting paid (and so well, by my Georgia minimum-wage standards) to simply love and look after the neighbors' dogs, all while being housed for two weeks, felt like the best kind of good luck. I took it as an omen that this move would work out just fine.

My charges were Blondie, an energetic blond terrier mix, and Buster, an ancient black Lab mix. Their owners were both successful interior designers and the kind of people made infinitely more attractive by their abundant warmth and welcoming nature. I liked them immediately and pledged to take better care of their dogs than anyone ever had.

For someone feeling displaced and homesick, as I was, the enthusiastic and unconditional adoration of these two dogs was a balm for my aching heart. I couldn't remember ever meeting two such faithful and affectionate animals, other than—of course— Biscuit, who was officially The Best Dog in the World.

Biscuit was a big fluffy white mutt, descended from a neighborhood retriever and her erstwhile poodle mate and brought into our family when I was still a toddler. We grew up together, Biscuit and I. She was my number one sidekick when it came to playing make-believe, dress-up, tag, and let's-sleep-in-the-yard-on-a-big-blanket. My older sister and I were raised Quaker, and part of that upbringing was an hour of quiet time a day, in which we had to play separately and quietly in our respective bedrooms or outside. She would host a tea party for her stuffed animals, or play schoolteacher to her doll students in her bedroom. Outside, I would put headbands and scarves on the ever-tolerant dog and dance about her, singing her special song that I'd composed: "Queen Biscuit, Queen Biscuit. Queeeeeeen of the Wooooorrrrlllld!" When I played my Madonna cassette tapes, I'd lift her paws onto my shoulders in an approximation of dancing. Often, she just sat quietly by my side while I worked on the stories and poems I loved to write in any one of my many collected journals and notebooks. She was undoubtedly my best friend right up until her sudden death when she was ten and I was twelve.

Sweet and affectionate as they were, these dogs also had an

impressively detailed health history, Buster in particular, and I had plenty of instructions to follow for their daily care. Buster was going blind from cataracts, had to take antidepressants for separation anxiety, and was on a strict regimen of Glucosamine and Chondroitin for his advanced arthritis. He was also completely deaf. He and Blondie were both on a diet of boiled chicken and rice; Blondie because she had a tender stomach, and Buster because he deserved the good stuff in his twilight years. Their water came from the Brita pitcher on the counter.

Biscuit had shared her Alpo with the rats that ventured out of our heavily wooded backyard, and her water came from the garden hose. The only medication she took staved off heartworms, which was standard for outdoor dogs. We hid the foul-smelling pill in cheddar cheese, which she gobbled with enthusiasm. Beyond her daily dog bone, she didn't enjoy many luxuries. She'd always seemed perfectly content with her lot in life as an exclusively outdoor dog. In fact, she wouldn't come inside the house even when invited. I can't account for why she was so averse to being indoors—she'd been that way for as long as I could remember. According to my mom, she'd come home to an unlocked front door soon after we got the dog. She bodily dragged a reluctant Biscuit over the threshold, saying, "Come on, Butch, go get 'em," in an effort to scare any home intruders that might be lurking within. The minute she released her grip on Biscuit's collar, the dog dashed for the door, eager to get out of the house and back to the yard, where she was happiest. Nothing about Biscuit suggested that she'd ever be a guard dog, an indoor dog, or butch in any way at all.

Though I'd long been passionate about animals and preferred their company to that of my own kind, I hadn't pursued an education that would lead to a career in animal care. I'd attended one of the top veterinary universities in the country, but I wasn't at all scientifically or mathematically minded, and I didn't have the stomach for the grislier side of veterinary sciences, zoology, or other related vocations. So I went with a clean and completely cerebral literature major, relegating my interest in animals to the extracurricular.

Now that I'd matriculated and was seeking employment in the real world, so far beyond the borders of my college town, I felt sure that my enthusiasm for animals coupled with my three years at the pet store more than qualified me to be the lady in scrubs who ferried animals from waiting room to private exam room, to weigh the pets and take their temperatures. Without overthinking my resolution too much, I applied for vet tech positions at twenty-five or so local animal hospitals.

I got one interview.

It was far away, at least by my intown Atlanta standards where everything—everything—is only a five-minute drive, except the airport, which takes fifteen. The thirty-minute drive to Hayward brought me to an unremarkable cement building along a suburban thoroughfare, where I met with an endearingly overweight guy named Andy. He left me in a sterile white seating area to fill out some paperwork, and, within a few minutes, it was abundantly clear that I was nowhere near qualified for the job.

Had I administered subcutaneous medications? Declawed cats? Neutered animals? Performed any anesthetizations? Diagnosed any illnesses?

I tried to glamorize my skill set—which was woefully limited to deworming (dumping the writhing masses into the disposal), administering medication (shoving pills down slimy, gagging dog and cat gullets), light medical attentions (applying shiny blue or pink plastic cat-claw tips; holding down a rabbit while my boss drained an abscess), experience with exotic animals (watching in horror as a monitor lizard took a rat by the testicles and slammed him to death against the cage wall)—carefully sidestepping the fact that I'd been less of a technician and more of a lackey.

Andy let me down easy, saying he'd be in touch. I knew better than to expect a call back. Three years' experience as a pet store cave troll does not a résumé make, and I had enough sense to spot the rejection between the lines.

While I was away from the dogs during the day, I was supposed to leave the jazz station playing on the stereo. This allegedly calmed Buster's separation anxiety, though I couldn't understand how that reconciled with his deafness. If he couldn't hear the music, was it the vibration of the jazz that he benefited from?

Buster's combination of ailments made him clingy in a very dear way. He kept me in sight at all times, even shuffling after me when I went into the bathroom. It felt good to be minded, to be needed. When I left the house, I took very seriously the owners' routine for reassuring him that I'd return. At the front door, I'd get down on face level with him and—because he couldn't hear me—I smiled and nodded exaggeratedly, petting him and kissing him, and then repeating the smile, nod, pet.

The smooth jazz was quietly thrumming when I returned from my failed interview, and I gratefully submitted myself to a session of pet therapy on the floor of the living room. Their slobbery approval was the perfect antidote to my slightly stung pride and growing anxiety over my lack of a professional Plan B. Somehow, Buster's gift of his stuffed squirrel deposited in my lap, followed by a sincere and thorough licking of my hands and arms, made it all feel less scary and uncertain.

That night and every night, I strapped Buster into a hunter green fleece-lined harness and hauled him up the stairs to the master bedroom, his limbs flailing and his toenails scrabbling helplessly for purchase. The idea was that I'd hoist him vertically, taking enough weight off his aged joints that he could go through the motions of mounting the steps. Only after we'd made it to the second floor, both panting—and, I imagined, equally relieved that the ordeal was over—would Blondie bound up the stairs to join us.

In the kennel, I frequently had to retrieve or deposit large dogs—some in excess of fifty pounds—to and from their fluorescently lit cubbies for walks or playtime with an interested customer. We ran an adoption program for local shelter dogs, and many of these were full-grown and slightly overweight, like Chester the resident Shar-Pei. He was adorable, but hoisting him or any of the other bigger dogs from the cage, and always holding tight to keep them from bolting for the exits until I could get them leashed, was a struggle. The squirmy dogs, so excited to be released from their claustrophobic enclosures, left me with back twinges that lasted for days.

But that was nothing compared to heaving the dead weight of an overweight arthritic Lab up a flight of stairs. Waiting for me in the bedroom, though, was a Sleep Number bed: the perfect place to collapse after my exertions. One side was extra firm, the other moderately so. I had written down the respective numbers on a bit of paper to be sure I could restore the mattress to its original settings at the end of my stay. Until then, I was bound and determined to find my personal number. I had been working my way through the thirties and was feeling almost close to perfect at thirty-nine.

Buster slept between the dressers, his toenails scratching erratically against the hardwood floor as he dreamed. Blondie's spot was right next to the bed, her satisfied-sounding exhalations the last thing I heard as I drifted to sleep.

After following up on all of my applications and accepting that the vet tech path was not to be, I started looking for openings at local pet supply stores. If I couldn't be the vet tech waiting room lady, I certainly had the chops to wash dogs and offer their owners advice on accessories.

True, working the sales floor at the pet store peddling merchandise—and pets, of course—had never been my strong suit. The pet counselors worked on commission, and they were ruthless, using every trick in the book to make a sale. This meant pitching all manner of questionably useful accessories (The pooper-scooper with ergonomically angled claw! Frilly underwear and pads for when your bitch is in estrus!) and upselling customers on industrial-sized bags of dog food, cat litter, aquarium pebbles, and so on. It also meant that many an $800 dog went home with the wrong family. And many of those dogs got returned within a week or a month when the first-time owner realized what they'd gotten themselves into.

We had one customer who kept his Siberian husky in the cab of his big rig while he drove around the country. I have never seen a more neurotic animal, or one more badly in need of exercise. The overwrought dog couldn't come into the store, which he and his owner visited anytime they were in the area, without chaos erupting. Within moments of the dog rearing his way through the cheerfully jangling door, displays were upended, toys and treats scattered across the industrial carpet, ferrets terrorized, the top layer of pig ears in the bin test-licked, and all other customers with or without their pets in tow hastened to the exit.

I didn't have the ambition to wheel and deal like the others, convincing new or expecting parents that a Lab puppy was a good choice, or that the twice-as-expensive memory foam dog bed was that much better than the regular and reasonably priced one. I was far more comfortable trimming toenails and chatting with customers about the consistency of their dog's barf than trying to convince them they needed a $150 tartan cushion to go with those nail clippers. Hence my permanent position back in the kennel with the animals.

Though working at PetSmart, Petco, Pet Food Express, or any other area chain felt like a big step down from being a vet tech, in both pay and prestige, I still preferred the access to the cute and furry that the job would provide over being, say, a barista or finding a desk job. I'd so much rather spend my days interacting with animals and their accessories than serving coffee to undercaffeinated customers or staring at a computer screen.

On my last day with Blondie and Buster, I was still without a job, or even a likely prospect. I was grateful that I at least had a place to stay, as Annie had invited me to use their spare attic bedroom indefinitely. We had worked out an agreement in which I would help ferry her boys to and from school, kung fu, tutoring, and so on, as well as do some light shopping and cooking in exchange for room and board. They rarely used their old Volvo, a car they'd been meaning to trade in for months, and essentially handed over the keys. Even with the question of employment still looming, at least I could check "roof over my head," and, for the time being at least, "set of wheels," off the list of required components for my West-Coast attempt at adulthood.

I compiled photos I'd taken of Blondie and Buster into a collage with dialogue bubbles declaring how much they'd missed their owners, and how much fun they'd had with me in the meantime. I left it propped on the kitchen island against Buster's economy-sized pill bottles.

It was only after the neighbors returned that I learned Blondie and Buster had regular dog walkers, a husband and wife team who lived a few streets over. They didn't offer overnight pet care, and they were looking to contract with someone who could provide this service for their clients, just as I'd been doing for the past couple of weeks.

I can see, in retrospect, why I was such an appealing solution to their problem. I showed up for my interview at Tom and Patty's house, a classic northern California bungalow with an adobe-shingled roof and fruit trees dominating the postage stamp of a front lawn, wearing Crocs and my favorite Big Smith overalls. When they opened the door, I greeted their two massive German shepherds first, kneeling before them to receive their exploratory sniffs and licks.

Tom and Patty later confided that this instinct on my part, greeting the dogs first, was more important to them than anything else I did or said in the hour-long interview that followed. In that meeting, conducted on their Southwestern-patterned couches in the dimly lit living room, I learned about their decade-old business, founded after they quit the rat race of corporate America to pursue their passion for animals. They shared with me the basic requirements of setting up shop as a professional pet-care provider, which, as small business ownership goes, had fairly low overhead. They primed me on which services were in high demand and the going rates for each, and where I might fit into this rapidly growing industry. Apparently, while plenty of pet-care providers would do end-of-the-day visits to tuck their clients' dogs and cats in, next to no one stayed the night. It was this highly sought service that would be my niche: the sleepover.

Of course, I would supplement these overnight stays with daily walks, helping lighten the load of Tom and Patty's packed client roster by picking up those neighborhood walks and drop-in pet-sitting visits they couldn't get to. Tom specialized in the group off-leash walks, so it was primarily Patty's portion of the daily walks I'd help with. I was not interested in—or rather, I was completely daunted by—the prospect of managing five or so off-leash dogs at once. They agreed that, as a beginner, I was better suited for the leashed walks with one or two dogs at a time. They'd have primary contact with their clients and would manage the billing; I'd do the work, taking home a contractor's percentage of what the client paid.

Beyond the business license, liability insurance, and a small inventory of basic supplies, I would need a few clients of my own. For tax reasons—to distinguish my role from that of an employee—my individually established business needed to have an altogether independent client list. This took a touch of extra explaining, as I was a comparative literature major to their combined double MBAs. The immediate takeaway, though, was that subcontracting meant that I could jump right in to working with them while figuring out my own marketing strategy and ramping up my business in parallel.

I had my marching orders, and an all-new plan.

Beyond my sheer delight at the prospect of spending my days, and many nights as well, in the company of so many different, affection-seeking animals, I was also enthusiastic about the business side of things. I'd always loved playing secretary, hoarding various types of ledgers and notepads. I got an adding machine for my eighth birthday and loved nothing more than accompanying my mom to Office Depot. Where most kids threw fits over candy or Barbie dolls, I'd beg for inane office supplies like carbon copies or the pink "While you were out" pads. For the life of me, I couldn't understand why my mom wouldn't let me play with her checkbook.

That I'd need to make business cards, track my mileage and gas expenses, file all work-related receipts for tax purposes, and submit invoices at month's end all sounded like too much fun.

In the very beginning, I only had one client I could call my own. He was Chase, a four-month-old cockapoo, which is a terrible name for a cocker spaniel/poodle mix. Chase's family lived in the neighborhood and knew Annie's boys through school. While Chase was undeniably adorable, with the giant heart-melting black eyes of a puppy and floppy brindle fur that was silky soft, he also had razor-sharp puppy teeth that he used early and often on the kids. According to the owners, he loved nothing more than a hand or an ankle to gnaw on—the younger, the better. His tendency to bite and chew extended to electrical cords, shoes, rugs, and anything else that came within reasonable range of his insatiable little mouth. Understandably worried about an accidentally electrocuted pup, and with the rate at which they were going through Band-Aids, they needed help training that puppy instinct to chomp, maim, and destroy right out of him. They were also very interested in the idea that I could leash train him and help teach him some of the more basic commands: come, stay, sit, no.

Because Chase was still brand-new to the leash, our walks were short. I did my best to teach him to heel and was careful not to pull back on the leash as he strained his way forward down the sidewalk. A firm believer in positive reinforcement, I came armed with ample edible rewards to dissuade him from asphyxiating himself on the other end of the leash. He was still so little, it was easier to simply scoop him up and address him face-to-face. But not especially effective.

The thirty-minute visit was filled out with backyard time, where Chase alternately attacked the fallen oranges from the citrus tree and nipped at my shoes. I threw a tennis ball as big as his tiny head from the safety of my perch on the picnic bench, my feet tucked up and out of range.

He was excellent at the chasing and biting part of fetch, but less so at returning his prize. I had to risk my fingertips as I wrested the ball away from him. If his sharp little teeth hurt that much when I played with him, there must have been some tears from the kids.

I was relieved to see some progress with the positive

enforcement training, especially with the "sit" and "leave it" commands. While I held a treat aloft, he'd settle on his haunches, the ball momentarily abandoned at his feet. Head cocked, he followed the morsel with his bright black eyes.

Unfortunately for me, when I praised him and offered him the snack, he chomped down on the treat and my fingertip together. At least I was able to sneak the ball away from him with my other hand.

After my visit with Chase, I continued on to clients of Tom and Patty's in the neighborhood. Thus far, they were all daytime walks or pet-sitting visits, as I had yet to start any overnight stays. I was grateful for the intro course of regular walks and house visits before what I felt was the advanced level of spend-the-nights, relishing the new rhythm of my days with the assortment of dogs I saw.

Next on the list was a favorite of mine. Pearl was a golden retriever and poodle mix, a golden doodle, just like my Biscuit was. Of course, back then, Biscuit was just a mutt with the smarts of the poodle and the loyalty of the retriever. Breeders have since caught on to the exceptional blend of qualities, and this mix has become a boutique breed with a hefty price tag attached. Where our dog was free, offloaded on our delighted family by a grateful neighbor whose retriever got knocked up by a local poodle, I am sure that Pearl cost a couple hundred bucks and was the result of a planned coupling.

Pearl looked so much like Biscuit, long-legged and fluffy, her curly cream-colored fur hanging down in her big brown eyes. While she retained the notorious gentle and sociable nature of the mix, she didn't quite live up to the reputation of golden doodles for being exceptionally smart and thus highly trainable. She struck me as being on the slower end of the spectrum when it came to intelligence, though she more than made up for this with her inherent sweetness.

And, though I'd never met them, her owners displayed some evidence of that, too. I could not figure out why they got a dog when they did. And this dog, too, needful as she was for interaction with and stimulation from her owners. Pearl stayed crated all day in the dining room and then had to wear a cone around her neck because she gnawed on herself, which they never took the time to train out of her when she was a puppy. At two, she needed exercise and contact—lots of both—and it hurt my heart to see her cooped up in that crate all day, every day. The chewing was probably just as much a symptom of boredom as anything else.

"Hi, cutes," I called to her as I entered, propping the Plexiglas door with my foot as I worked my key out of the old lock. I was quickly finding commonalities between the houses in the area, one being that the locks were old and the doors even older. Many of the houses were split-levels, renovations and updates seemed infrequent, and there didn't appear to be much turnover. A *For Sale* sign was a rare sight indeed. People came and settled along the wide tree-lined boulevards, raised their families, and there they stayed for decades to come with seemingly little need for upgrades.

This family seemed to fit that pattern. From what I'd gathered, they had a little boy—blond curls and smiles from the pictures on the mantle—and there was another on the way, based on the furniture reorganization going on upstairs. I only went up that narrow staircase when I needed to use the bathroom. I tried to avoid that, though, because the toilet ran, and once I almost broke off the door handle from the inside. Getting stuck in a client's bathroom would surely be bad for business.

Pearl's crate was in the open living room/dining room to the right of the front door, but it was situated behind the dining room table, so I couldn't see her until I was almost upon her. As was her usual, she sat facing the crate door, the cone half obscuring her adorable face. Her expressive eyes were barely visible behind her untrimmed doggy bangs. She wagged her tail, a *thump thump*

thump on the base of her crate that was one of the most comforting sounds in the world to me.

But I had to be ready. The minute I sprang the latch on her crate, she'd be out and past me faster than I could ever get used to—I'd been clipped by that cone more times than I cared to recall. She got so excited about her release that she often peed a little on the floor before she made it out the back door to the yard.

I was slowly learning some tricks, though. It appeared that I was trainable, too. But, like Pearl, maybe not exceptionally so.

"Be right back," I said to her.

She whined a little as I dashed down the stairs from the kitchen, through the playroom, and to the back door, which I unlocked and opened in preparation for her mad dash.

It was a good move, too, because the minute Pearl made it beyond the threshold, she squatted. She wasn't even on the grass when she let loose a stream on the paving stone right outside the back door. Not ideal, but far better than the playroom floor.

When she finished, she shook her butt in what I read as delighted relief and then trotted over to me, trying to lick my hand and instead mauling my thigh with the hard edge of her cone. I removed it for our visit so that we could have some productive ball time, and I spent a good while scratching her neck where the collar had been.

For the next thirty minutes, we threw the tennis ball and I taught her about sitting and waiting for the ball toss. It took her longer than most dogs to get the hang of it, but by the end, she was returning to me with the ball in her mouth, if not yet dropping it at my feet. After replacing the e-collar, and before returning her to the crate, I was sure to give her a prolonged full body scratch—an apology of sorts for subjecting her once more to the cone of shame and then locking her up. That was a mean one-two combination.

She moaned a little, a low guttural sound that Biscuit used to make during belly rubs. Doggie ecstasy. She tried to lick my hands and legs in gratitude but just further bruised me with hard plastic. Pants were probably a good idea for my visits with her, in spite of the warm Indian summer sun that made me sweat and her pant.

My first overnight gig was in less than a week. The assignment was a five-night sojourn with two dogs and seven exotic birds. In the meantime, I'd secured my business license and liability insurance and had filed my DBA with the local paper. I'd also joined an association of pet sitters and dog walkers. They were in need of a secretary to take down the minutes at the monthly meetings, so I'd offered my services there as well. I had business cards, a new email address, and an established scale of rates for my various services. But I had little idea of how to prepare myself for such a complicated client as this, with so many different breeds of birds, each with specific dietary and environmental needs. I had limited experience with birds of any kind as it was, much less with the large talking variety. Ready or not, though, I was poised to become a professional purveyor of the pet slumber party.

Journal entry: Tuesday, 8:00 AM

Looks like all that note-taking during episodes of
Head of the Class and all those sick days spent playing
secretary are finally going to pay off! I can now
officially add Secretary, capital *S*, to my title—the
kind that sits on a board, though, not behind a desk.
I'll be taking down and presenting the minutes at each
meeting of the local pet-sitters' association, and I get
to include official-sounding things like, "Respectfully
submitted." Just like Kinsey Millhone! Dream realized.
My pencils are sharpened and my new notebook is
ready for tonight's meeting. Maybe I should get a
clipboard, too?

CHAPTER TWO

Wild Birds of the East Bay

....

T he bungalow, with its terra-cotta tiled roof and wide front porch, had a deliberately wild-looking front garden. Fruit trees and bright colors dominated the otherwise tidy space, the recently mowed lawn enclosed by a low, natural wood fence. Still unused to the typical Californian abundance of produce hanging at eye level, so different from the soaring oaks in the south of my childhood, I resisted the urge to pluck a lemon from the branch nearest me before ascending the front steps.

What I knew before meeting Bev: she had seven birds—two cockatiels, two parrots, two budgies, and a conure—plus two vizslas, Hungarian hunting dogs. She was going to Bermuda for her daughter's wedding.

When Bev opened the door, my impression was midsixties, beautiful, and braless. I was even more impressed by how completely the house was given over to the animals. Birdcages lined the open-plan living room and dining room like furniture, and the couches and chairs were covered in dog blankets and the dogs themselves: lean, with sleek brick-colored coats, mustard yellow eyes, and floppy, velvety looking ears. At only thirty-five pounds or so, their short fur and lithe little bodies conveyed strength and speed. They regarded me with comfortable indifference. Guard dogs they were not.

Bev offered me tea, and a mammoth binder filled with full-color bird biographies, each in its own plastic sleeve. We took a tour of the birds while sipping pungent Lapsang Souchong that tasted like a campfire.

We started in the office in the back corner of the house, which opened out onto an expansive back patio. This was the territory of the budgies—palm-sized, neon-colored noisemakers, from what I'd experienced of these parakeets in the past. They were louder than they were large. According to his profile, Echo—electric green to Bindi's more muted blue—was a biter. He and Bindi shared the office with Nora, a blue-crowned conure.

I can't say I'd ever seen a conure before, or, if I had, I didn't realize I was looking at one. Nora was emerald green up to her neck, where the blue crown took over, her head fully feathered in aquamarine.

"She responds to 'Be quiet!' Or you can just put a sheet over her cage." As if on cue, Nora emitted a screech that I feared might sonicate my bones. Bev looked entirely unruffled, while I was pretty sure I'd peed in my pants a little.

In the middle room, which was more of a hallway between the back of the house and the kitchen, lived Aphrodite and Sterling, both cockatiels.

"Aphrodite curses in Spanish. And she is the loudest." I felt certain that it was scientifically impossible for her to be any louder than Nora, but I nodded.

"She also replies to 'Be quiet,' but she may mouth off at you," Bev said with a smirk.

I rested my tea on a shelf in the walk-through to free up my hand to take some notes of my own. *Nora: holy shit, loud. Aphrodite swears.*

In college, my thesis advisor had a talking parrot named Liebling, an African gray. At every meeting, Dr. James proudly showed me the updated list of words that she'd taught the bird.

I was writing my thesis on the nonfiction works of Barbara Kingsolver. I fancied myself an ardent nature- and animal-lover and was an enthusiastic captive audience when Dr. James got talking about her pets. In addition to Liebling, she had two outrageously fluffy American Eskimos. By the time I'd submitted the final draft of my thesis, I was addressing her as Sarah, and she was emailing me pictures of the dogs.

My mom, a linguist by training and an audiologist by profession before she ever got into teaching pronunciation, has always been fascinated by animals' ability to speak. When she was a student, she studied apes' ability to sign and communicate with pictures. So whenever my thesis or meetings with my advisor had come up in conversation, she'd always asked about Liebling.

I was already anticipating that this gig would yield plenty of talking-bird gems with which to regale her. And it made me wish I'd kept in touch with Dr. James. Sarah. She'd have surely loved hearing about my talking avian charges and would've wanted a complete list of their vocabulary.

"Sterling is a handful," Bev continued. "He has to be left out of the cage during the day, but he's a big chewer. He's afraid of brooms, so if you leave it propped against the door jamb, he won't go into the kitchen." She gestured toward the broom that we'd passed under as we came into the back of the house.

It was only then that I noticed the extensive damage to the molding, and what little was left of the windowsills. I could see that the woodwork around the windows and doorframes had once been quite lovely—intricate, even—but had since been gnawed to splinters by one, or perhaps a few, very sharp beaks. Just about everything I'd seen so far looked chewed-on, actually. Chunks of wood were scattered across the floor, and they crunched softly underfoot.

"Make sure Aphrodite's cage is flush with the wall and that there are plenty of toys attached to her door, or Sterling will break into her cage."

"Break in?"

"Yeah, he knows how the door works, so he can open it and let her out. The toys distract him. *Sterling,* I wrote, *B&E—watch out!*

"This," she said, turning away from the cages and toward the sideboard, "is the fish. Just a few flakes a day, and he's fine." *The Fish, few flakes 1x/day.* From what I could tell, he didn't get a page in the massive pet info encyclopedia I was hefting. Or a name, for that matter.

Ducking beneath the broom, Bev led the way through the kitchen, her gauzy tunic trailing behind her. "And then in the front of the house, we have Krishna and Bonsai." I checked the binder and saw that Krishna was a Red-lored Amazon, Bonsai an African gray, like Sarah's Leibling.

I was trying to do a quick calculation of how many continents these birds and dogs collectively spanned. Australia, Central and South America, Africa, Eastern Europe. With Bev's Asian sensibilities, this was a decidedly worldly household.

Like Nora the conure, Krishna was also bright green in the

body, but her face was brightly accented with red and orange around the polished knuckle of her beak.

"Krishna has a broken wing. She'll come out to sit on your finger and have a pistachio or a cashew, but she doesn't like to be petted."

I was scribbling furiously as we moved to the last cage, occupied by the surprisingly small and comparatively dull gray parrot.

"Bonsai is a plucker, so he wears this bell to distract him."

Bonsai jingled as he shifted from one foot to the other, seeming to respond to his name. The bell that hung from his neck looked like an ascot, giving him the dashing air of a dandy. All he needed was a little bird-sized bowler hat.

"He and Sterling don't get on so well, but Bonsai can always hide in the towel hutch—I'll show you in a sec." We hadn't even gotten to the dogs and their schedule and preferences yet, and I'd already taken two pages' worth of notes.

The dogs hadn't moved since I arrived. I assumed Sasha was the smaller of the two. She was curled in a recliner in the corner by the fireplace. Max, sprawled out on the love seat, raised his head from his paws as we entered the living room.

"The dogs can't take the sun salutation, so they need to go out for their walk." I discreetly scanned the binder for word on the sun salutation. My mind was going to the yogic practice, which seemed to fit with this woman's aesthetic. I couldn't figure how that applied to the dogs. "Otherwise they go crazy, barking and running around. The birds are loud enough as it is without the dogs' howling along."

Ah, the birds. The birds' sun salutation. I was no expert, but I felt pretty sure this was when the birds greeted the rising and setting sun with prolonged squawking. No doubt I'd find a thorough explanation somewhere in the tome I carried.

We headed upstairs to see where the dogs and I would be sleeping. At the top of the steps was a narrow wooden hallway with two rooms to the left and a bathroom at the end. Bev stepped through the first doorway. If I'd expected dog beds or cushions, there were none. It was a small room with two walls of windows, no curtains, and an unmade king-sized bed with mismatched sheets, the flat a plaid and the top sheet a loud Hawaiian print.

"Sasha will sleep up with you. She's a lovebug. And Max'll sleep down at the foot of the bed."

I'd never slept with a dog in a bed, and it sounded kind of lovely. I'd have enthusiastically slept with Biscuit, but, given her inexplicable insistence on remaining outside, this meant infrequent naps with her on a quilt spread out in the yard. Even during the storms that scared her so much, she would only come so far as the basement or the screened-in porch, where she'd cower on her tattered blanket, shaking visibly at each thunderclap.

Bev exited into the hall, and I followed, catching a glimpse of the room adjacent, which seemed to be filled from floor to ceiling with piles of crap—boxes, clothes, paper, and towering stacks of books.

As Bev descended the stairs, she said over her shoulder, "So, do you have any questions?" I mentally scanned the massive information overload of the last half hour or so. Who curses in which language? Who likes which nuts cracked partially, versus hardly at all, versus completely shelled? Had I seen that one of the birds likes buttered toast?

"I think I have everything I need," I said, patting the binder with more conviction than I was feeling.

"Great. Then here is the key. You have all the contact numbers and emergency info. And plenty of reading material on these guys."

Bev didn't own a coffeemaker.

I discovered this on my first morning. I'd made the mistake of pulling the blankets off the birds' cages as soon as I got downstairs, not fully grasping that this would signal them to wake up at once. Wake up and start cawing.

The dogs were ignoring their breakfast entirely and howling so loudly that I could actually hear them amidst the window-rattling, earth-shaking cacophony of bird calls that was making it hard for me to keep my eyes open and my fingers out of my ears.

I hadn't slept well the night before, no doubt because I was not at all used to sharing a bed—forget the pillow—with dogs. When I'd gone up to the bedroom to turn in for the night, the bed looked just the same as when Bev had showed it to me—same sheets, unmade, Max's nest still intact at the foot of the bed. I couldn't help but feel a little weird about sleeping in already-slept-in sheets. I'd made a mental note to find the washing machine the next morning. Then Sasha had wriggled her way under the covers with me and put her head down on one half of the pillow with a contented-sounding sigh.

Even with Sasha fully tucked in, using the pillow just like a human, she wasn't as cuddly as I'd expected her to be. She didn't invite a snuggle at all, but put her legs out straight in front of her so that she occupied the greater half of the bed. With her at my side and Max at my feet, I'd turned out the light. I found myself holding my breath a little, lying stiff and unmoving, in the same way I did when I shared a bed with another person. Like my breathing, or the slightest movement, would disrupt their sleep.

Contrary to the cozy comfort that I'd anticipated in sleeping with these dogs, the reality was that Sasha was a total pillow hog, unselfconsciously exhaling her meaty-smelling dog breath all over my face in the night. When I turned my back to her, seeking out a corner of the bed that I could claim as my own, she also proved expert at kicking me while she slept, the blows landing right in my left kidney.

In the morning, bleary eyed and a little sore, I woke to find myself nose-to-snout with Sasha, her angular head still on my pillow. Max was awake and inching his way up the bed toward us. It wasn't until I moved to rub his belly that I saw the shiny pink of his raging erection. I threw the covers back without too much regret, in spite of the early hour.

Standing in the kitchen without coffee, wearing my saggy pajamas pants and no bra, listening to a cockatiel demand toast of me, I caved and took the dogs out. They shot from the front door as though I'd packed them into a cannon with a steel rod. I could've sworn there was smoke coming off Max as he landed in the yard, never even touching the front steps on his way down.

I crossed my arms protectively over my unsupported chest, and we ambled down the sidewalk, the dogs growing calmer as we gained distance from the house. They happily wagged their stumpy tails, along with the whole back half of their muscular bodies, sniffing every fence post, shrub, and telephone pole. I counted the number of houses we passed until I could barely hear the birds any longer. Three, four, five. I heard a call that was higher and shriller than the others and wondered if that was in fact Aphrodite, louder than the rest after all. Seven, eight, nine.

Inside the house, the noise had been so frenzied and confusing it was impossible to sort one bird's call from the next. The further away I got, the more distinct each caw and cry became. By the eleventh house, I couldn't hear them unless I stopped walking and strained really hard. We were nearing the public park at the end of the street, so we crossed, and the dogs took off into the outfield of the baseball diamond. I hadn't brought a ball or a toy, but I found an adequate stick and gave it a toss. They didn't even notice I'd thrown it, even though it fell a few feet from where Sasha was bounding about. Clearly not retrievers.

Back at the house, the vizlas flopped down on the furniture. Sterling was still asking for toast. And there was still no coffee. I should have paid better attention to the implications when Bev had served me tea from her vast collection. Definitely not a coffee drinker.

From my extensive reading, I understood that each bird had a very specific diet. After putting two pieces of toast in the toaster oven—one for me, one for Sterling—I got down to sorting the requirements for each. There were apples to chop into specific sizes, supplemented by mango and papaya, as well as some broccoli and lettuce. There was an astonishing assortment of nuts to crack in varying degrees, meant to supplement each breed's seed blend, mixed with a handful of ZuPreem fruit pellets and just a sprinkle of something called Tropimix.

I started in the back room with the two smallest birds, dumping their uneaten pellet mix and nuts and fruits and veggies into the compost bin. According to the instructions, they got no more than two cashews, a couple of peanuts, and maybe a pistachio as a treat.

Since Echo was allegedly a nipper, I was careful to get in and out of his cage quickly. Nora took her walnuts partially cracked, which she received without a peep, to my great relief. In the hallway between office and kitchen, I fed the fish first, if only to spite Sterling, who was still crowing about his toast. As a matter of principal I ate my piece first, which didn't sit well with him at all. Even if he did say, "Please," I wasn't eating second to a bird.

After Aphrodite got her blend of pellets, broccoli, lettuce, and fruits, along with her peanuts (in the shell, un-cracked), I moved on to Sterling's breakfast. He took the toast in his beak, finally shutting up and focusing instead on tearing impressively large chunks out of the buttered bread while maneuvering it with his talons. He received double the food: bowls inside the cage and on the top as well, near his exterior perch. His favorite nut was the walnut, which I lightly cracked on the butcher block with a hammer.

Krishna ate shelled peanuts, while Bonsai preferred his in the shell ("He loves a challenge!" according to his printout), so I only cracked his nuts lightly with the handle of the hammer rather than the head.

Bev had asked me to speak to the birds. This shouldn't have felt as strange as it did, since I'd always spoken to dogs and cats— those with whom I'd grown up, cared for at the pet store, and even met on the street. I think the primary difference was that not a single one of those animals had ever spoken back to me, as Aphrodite and Sterling could. I'd never been spoken to by a bird— or any other non-human—before, and it was far more off-putting than I'd anticipated.

"All right, Bonsai," I ventured. "Yum, yum. I like your bell!" He, unlike his more fluent friends, said nothing.

With the riotous sun salutation behind us, and the birds happily sated on their nuts and seeds and fruit, it was time to start back at the beginning with Bindi, Echo, and Nora's cages. This time I washed and refilled their water bowls and replaced the food- and droppings-spotted newspaper lining.

I hate the feel of newspaper; I have for as long as I can remember. When I was young, recycling was my household chore ("If you're going to be part of this family you have to contribute!" my parents said). I shuddered to handle the weeks' worth of *The Atlanta Journal-Constitution* and *The New York Times*, packing them in to paper bags to be placed curbside. I can only guess that my loathing of that smooth, sooty texture stemmed from the moment when, at the inquisitive age of four, I decided that I'd like to know what a cardboard box tastes like, so I licked one when no one was looking.

By the time I got to Bonsai's cage in the front room, my hands were so covered in bird pee, guano, and wet seed that I hardly noticed the feel of the newspapers anymore. I was effectively wearing bird-shit gloves. Cleaning Bonsai's cage was an exercise in balance, keeping all the peanut shells from tipping onto the hardwood floor. This, now that I looked more closely, might not have been such an issue after all, since there was plenty of poop splattered along the floorboards as well. I mentally added mopping to my list of morning to-dos.

After I finished with the cages and the sweeping, cleaned the kitchen of fruit rinds and shells, packed away all of the seed in the pantry, and mopped up the crusted remnants of bird droppings from the floor, I was feeling like Cinderella of the animal kingdom. Before I left Sasha and Max and their many feathered friends for the day, I went back through the binder to be extra sure I hadn't forgotten any critical steps or instructions. The dogs could access the back patio throughout the day by going through the kitchen, under the broom, and out the dog door in the back room. Sterling was roaming free, no doubt annoying the shit out of Aphrodite. I was pretty sure I'd heard a *"cabrón"* out of her, directed, I assumed, at her neighbor. Bonsai tinkled merrily as he worried his peanut shells. Nora had emitted more than her share of pee-in-my-pants squawks. As long as I was back before the sun started to set, they'd be fine for the day.

Though dark wouldn't fully descend until around eight o'clock, the birds' second sun salutation of the day began as soon as the blindingly golden light of late afternoon started slanting through the window blinds at a certain angle. They knew collectively, instinctively, that the sun was waning and it was time to celebrate. Walking down the sidewalk from where I'd parked, I could clearly hear the jungle concert in full force. I knew from that morning that the chorus shook the house. What must the neighbors think? I had yet to encounter anyone on the sidewalk during my walks with the dogs, but I was desperately curious to see their reaction to this twice-daily assault on the ears.

Opening the screen door and then the weathered wooden front door, I could also hear the dogs' frenzied growls and whines within. They turned tight circles in the living room while the birds made their joyful noise. With the front door open wide, Sasha and Max were out like ochre-furred bullets into the yard, turning larger and less-frenzied circles there until I opened the front gate out onto the sidewalk.

At least one of the daily walks had to be long and vigorous enough to even come close to tiring these guys out. If I did it right, we'd get back to the house as the sun set, casting long lavender shadows over the low houses of the neighborhood.

In Max and Sasha's neighborhood, the sky was just as wide open as anywhere else in the East Bay, and I found myself standing on the sidewalk with my head thrown back, tracking jetliners as they inched across the darkening sky. I was used to the soaring trees of my Atlanta home. The dense green growth there shielded us from the Southern sun and limited our exposure to what filtered through the canopy, dappling the landscape with a shadowy, shifting light. I was already addicted to the bright, unobstructed California sunshine that drenched everything beneath the perpetually cloudless crystalline sky.

I couldn't speak for the dogs, but I was certainly tired out after our walk. Between all of my other visits that day, I hadn't even walked that far—maybe five miles.

Before heading upstairs for the night, I placed blankets over the cages, returning Sterling to his cage last.

"Good night," I said.

"Good night," he replied.

It was barely nine o'clock, but I was ready to crawl into bed. I only hoped Sasha would lay off the kidney shots.

I arrived at the house on my third evening of bird duty to find Bonsai's cage spattered with red. The shells lining the bottom of the cage were flecked with white guano and the deep crimson of congealed bird blood. Sterling was perched high atop the armoire in the living room, looking like he'd swallowed a canary.

"Hello," he called to me.

"Fuck!"

The broom was lying across the kitchen floor, presumably dislodged by one of the dogs on their way to or from the back door. I extended my arm to Sterling, who turned his head demurely away.

"I'm not asking. Come. Here. Now!" I used both hands to grab him, and he gave a squawk. Once he was locked in his cage with the broom in its right place, I opened Bonsai's cage. He hopped nimbly up on my hand, favoring his left leg. His right leg was mangled, the blue gray of his skin torn and still bleeding. I'm not a bird expert, but I knew that birds do not have a lot of blood in their bodies to lose. At the pet store, we'd used a yellow powder to staunch the flow, in those rare instances when they had reason to bleed. This wasn't covered in these birds' notes, though, and I had no idea where to even start looking for a little bottle of coagulant.

I placed Bonsai back in the cage and went to the garage where the bird carriers were. Thankfully, Bev had noted this in the binder. Bonsai went into his Pet Sherpa without objection. I only dimly registered the additional damage Sterling had inflicted on the window molding during his rampage, a fresh dusting of wood shards and paint underfoot as I shuffled awkwardly out the front door and through the gate to the car, trying not to bump or jostle the unwieldy carrier too much.

I called Bev from the laminate-and-upholstered chair in the waiting room of the veterinarian's office. Bonsai was in the back, being examined. According to Bev's vacation itinerary, she should've been at her daughter's rehearsal dinner. Ashamed at my cowardice, I was hoping against hope that I'd be able to leave a message instead of having to explain the situation to Bev in real time.

My message was brief and to the point, and I asked her to please return my call at her earliest convenience so that I could update her on Bonsai's condition. Soon after I hung up the phone, the doctor emerged.

"He lost a lot of blood, but he's stable. We've cleaned and wrapped the leg and started him on antibiotics. He'll have to wear a collar to prevent him from interfering with the bandage."

"A collar?"

"Yes, a cone, around his neck."

"Ah."

"How are you at administering oral meds to a bird?"

"Oh, I am fair to . . . ya know . . . good." I'd never done it in my life.

"Are you ready for the bird, then?"

"Yep." The doctor must have sensed my hesitation, though, because he demonstrated how to squirt the yellow liquid down the bird's beak using a hand puppet as a stand-in for Bonsai.

Luckily, the vet agreed to bill Bev by mail so I didn't have to worry about the payment. The only substantial amount of money I had to my name was the graduation gift from my parents, sitting in a no-access, high-yield CD at Bank of America. Beyond that I had about $72.

I'd seen the charges, reading the final sum upside down as I stood at the counter waiting for the vet tech to bring out the bird. I shouldn't have been surprised, considering the leg cleaning and bandaging, the collar, antibiotics, and the demonstration on the puppet, which they surely charged for, too. But I was. Shocked, even. I wondered if $1,200 was par for the course when you have exotic birds for pets. But mostly I was just relieved to not be held responsible for the balance right then. Or hopefully ever, though I had no idea how Bev might react to the news and judge my culpability in the case of Sterling versus Bonsai.

Back in his carrier, Bonsai's cone kept scraping the top and sides of his plastic cage. His e-collar was a comically tiny version of the one that Pearl sported. I was impressed that the vet stocked cones that small, though perhaps it was about the right size for a Teacup Chihuahua or toy poodle. I had Bonsai's meds in a white bag with instructions tucked inside, just like I'd get when I picked up a prescription at the drugstore.

Once we were home, I left Bonsai in his carrier while I scrubbed his cage clean of blood. At last, I gingerly replaced him on the lowest, most substantial limb. Though his leg was wrapped, he still had use of his claw. He was holding it suspended above the perch, tucked close to his body in a way that made him look like a sleeping flamingo, but for the absurd-looking cone around his head. He seemed to be peering at me with a look of reproach.

Since finding Bonsai in his gore-flecked cage, I'd been fighting a growing anxiety that I'd completely missed—or misheard—Bev saying that Bonsai should be left out of his cage while I was away during the day. Only now that it was too late, I heard her voice in my head saying he'd fly into the towel hutch if Sterling should get past the broom barrier.

I reread Bonsai's write-up on the last page of the binder with some measure of dread at what I might discover. In all caps, written beneath his name and breed: *OUT during the day*. Idiot! In the event that Sterling slipped past his broom and attacked, Bonsai would fly to the towel hutch in the hall where he could hide from the larger, stronger bird.

I walked into the hall by the downstairs bathroom and flipped on the light. Sure enough, there was a telltale gray feather in the topmost stack of towels from the last time Bonsai had sought refuge there.

Without flight from his cage, poor Bonsai was a sitting duck. Sterling had grabbed him through the bars of the cage and pinned him there on the other side of the wire, mauling him repeatedly. Inside his cage, Bonsai wasn't safe. He was defenseless.

And I was screwed.

"I'm sorry, little buddy," I said to Bonsai. He continued to stare, unblinking. I walked back to Sterling's cage, maneuvering around the broom.

"You're an asshole." He fluffed his feathers and said nothing. "No toast for you tomorrow."

If there was any hint of a silver lining to Bonsai's injury, it was that I didn't have to shower with him. In the binder notes, I was instructed to take him into the shower with me and deflect some of the spray from the showerhead over his feathers. Far stranger than sleeping with dogs, showering with a parrot felt complicated in all sorts of stressful ways. How would I know what was enough, or too much, water? I didn't want to inadvertently waterboard the poor thing. On a very basic level, was it weird that I felt really . . . weird about being naked in the shower with a bird? However obvious this process might have seemed to Bev, it wasn't to me.

Now that Bonsai was coned and bandaged, I could've bagged

his leg securely against the water and temporarily removed his cone. But, no. I'd already broken Bonsai once; I couldn't bear to risk doing it again.

Bev called the next day as I was coming back into the house from the dogs' morning walk. My heart was in my throat, fully expecting some combination of rage and ridicule at my stupidity. I felt certain that I'd be responsible for the bill, too, which was totally fair.

Bev was thankful for what I'd done for Bonsai and utterly unfazed by the astronomical cost of the vet bill. Even after she'd said her part, I continued to explain my logic, assuming Bonsai was safe from Sterling in his cage and never realizing that the opposite could be true. Though I had read and reread the notes, my certainty that the cage was impregnable must've played tricks on my eyes and caused me to elide the obvious instructions.

"Of course I understand," she said. "Just keep Sterling and Bonsai caged until I am home." Either Bermuda was a magically restorative and calming place, and I had it to thank for the reprieve from a $1,200 punishment, or else braless Bev of the tea and the yoga really was that Zen and benevolent.

In my great relief and infinite kindness, I relented and fed Sterling his toast after all. Even standing at attention in the kitchen between Sterling's and Bonsai's cages, I wasn't taking any chances, not even allowing him to enjoy his breakfast in freedom. He took umbrage at being fed his toast while locked up, instead of on his usual cage-top perch.

"Sorry, fella," I said as I latched his cage after refreshing his water bowl. "You brought this on yourself."

Poor Bonsai couldn't manage his usual feeding routine with the unwieldy cone around his neck. He maneuvered well enough to reach his pellets, water, and fruit and vegetables, but the cone prevented him from holding his peanuts close enough to his beak for gnawing in the manner he was accustomed. I fed him the nuts, stripped of their shells, and a couple pistachios as an extra treat. Sure, I'm anthropomorphizing, but his resentment felt palpable. As he snatched each nut from my outstretched hand, he stared me down with an angry glint in his black eyes.

Before I left the house, I triple-checked the locks on Sterling's cage and left the broom in place, propped across the entrance to the kitchen. As little trust as I had in Sterling, I had even less in myself. Despite my diligent review and frequent referencing of the instructions provided, and my fastidious attention to detail when it came to feeding the birds and maintaining their cages, I'd still managed to screw up in spectacular fashion. As prepared as I'd thought I was, I wasn't nearly prepared enough.

I couldn't imagine that many pets would require care as involved or specific as the birds had, but that remained to be seen. In any case, the stakes would always be just as high. I sincerely hoped that even more careful review of my every action—and fewer easily enraged animals—might result in less violent sleepovers in the future.

Upon her return, Bev gifted me with a book on herbal medicine. So kind, and so very random. I secretly hoped that I might find the secret to Bev's outrageous magnanimity revealed within its pages.

After depositing my check, I updated my business profile on the pet-sitting association's website. Experience with exotic animals: check! Though there wasn't a field for it, I was mentally noting that I now had bona fide experience with animal-on-animal aggression, too.

Little did I know how indispensable this skill of managing the wilder and less-predictable aspects of the menagerie in my charge would prove. I'd leaped enthusiastically into the pet-care industry for the serenity, the simple joy, of spending my days in the company of animals. But my job, it would seem, was more about maintaining the illusion of control.

Hi Annie,

I have an overnight visit with the pug brothers scheduled for tonight, so I put dinner in the fridge. Greek chicken. Yum! I'm with these dogs through the weekend but should be able to have dinner with y'all at least a few nights this week. Just going early this first day to get the lay of the land.

Pugs! Chicken! Love you.

Lindsey

CHAPTER THREE

Nanny Cam

....

U ntil I started walking the Tervuren shepherds, or Tervs, I didn't know that there was such a thing as an agility dog. Or that these dogs competed against one another and won awards for their speed and precision. Hunting dogs, sporting dogs, herding dogs, guard dogs, lap dogs, sure. I'd watched Westminster. But agility dogs were all new to me.

There were three of them: Zipper, the female; Rascal, the male—both seasoned champions, judging by the statues and ribbons that crowded the mantel in the living room—and Slinky, the newest addition to the family and still "in training" through her puppyhood.

On their cul-de-sac, I parked next to the owners' giant passenger van bearing a Terv sticker on the bumper—the touring mobile, I assumed, for when they traveled to competitions. By the time I reached the front gate, Zipper was already barking through the mail slot. At the porch, the front door was shaking from the impact of her front paws. As I unlocked the screen door, I braced myself to make like these dogs and be as swift as possible. Once the front door was unlocked and open, I'd have to somehow be quicker and stronger than Zipper to get in and shut the door without her and Rascal bolting past me into the front yard. I had zero confidence that the waist-high white picket fence would hold them for a moment.

Contrary to my initial assumption, Belgian Tervurens and German shepherds have little in common. The Tervs have a sleeker build and a shorter stature. Their snouts can be so long and narrow as to evoke a collie's face. Their long, glossy fur is soft to the touch and fluffs out in a ruff around the head. While both breeds are highly trainable, intelligent, and most successful as pets if they are given a job like herding or guarding, the more time I spent with this trio of Tervs, the more I wondered at my own ignorance that the two could be confused or even compared.

The German shepherds I worked with were keen to please and get their job done with care and efficiency; the Tervs' sole focus seemed to be getting where they're going or obtaining what they want, and fast. Of course, that could have been specific to these particular dogs and not all Tervuren shepherds. I was quickly learning that for every generalization about breeds, there were as many exceptions that challenged my previously held conceptions. Environment and especially training seemed to have just as much to do with any dog's temperament as their genetic provenance.

The athleticism of these particular Tervs, at least, overrode any impulse control or command to slow down. Which is where the extensive agility training surely came in handy. Just as much as their speed and precision garnered them awards, it seemed that same conditioning to stop, sit, wait, and heel were equally beneficial for the humans—and any other dogs—that came into contact with them.

The front of the house was entirely overrun by the dogs. Toys were strewn over every surface; there were water bowls of varying sizes and heights in strategic locations throughout the entryway, living room, and kitchen. The pantry was stacked deep with treats for training, reward, and dietary supplement. Their owners administered these very carefully depending on the dog and the context. It took me at least a week's worth of visits to get the treat distribution down pat.

Then of course, there was the mantle of fame. The owners were going to have to extend the mantle or add a shelf, as the trophies, certificates, statuettes, and ribbons were threatening to spill over onto the floor. Or, more hazardously, into the fireplace itself.

Once the front door was bolted, offering no option of escape, Rascal backed off to see what I had to offer. Zipper was barking madly, running laps at warp-like speed around the large center kitchen island, while Rascal awaited my next move, his intelligent eyes trained on me and his head cocked. Though he was still, he was wired for action, his whole body tensed for my cue. He was significantly larger than his female counterpart, standing about half a head taller. I was grateful that it was the smaller of the adult dogs that was so excitable. If he were as wild as Zipper, there would be no managing him.

What I had to offer Rascal, always, was freedom from the house in the enclosed backyard. But first, I had to retrieve the third corner of this Terv triangle. Slinky was still being crate trained and was kept in the back bedroom in a handsome mahogany affair that had to have cost a bundle. The wooden slats of the cage matched the wood of the dresser and bed frame, her enclosure as much a piece of furniture as a tool for training. She waited patiently as I unhooked the latch to her crate, and she accepted my caresses and kisses upon the supernaturally soft crown of her head, offering me a reciprocal lick on my hand.

Though her face was still puppy-cute and her head and paws just slightly too big for her growing body, she already sported the lean, angular frame of a Terv. Unlike the older dogs, who cared singularly about getting out of the house, she was content to hang back a bit, keeping pace with me as we worked our way through the house. Whether it was the puppy in her, or the innate animal knowledge that she was the submissive in the pack—the low dog on the totem pole—Slinky was distinctly well-mannered. She was a little lady, deferring to others carefully in a way that came across as dainty.

As I picked my way across the living room minefield of rope toys, stuffed animals, kongs, and other miscellanea that may not have been intended as toys but fell victim to the dogs' jaws nevertheless, Rascal clued into my trajectory. As though a switch had been turned on, he joined Zipper's mania as they tore through the kitchen toward the back door.

I released them from the house-as-prison into the gravel alley just beside the back door, which the dogs had been trained to use as a bathroom. It smelled awful as ever in the warm midday sun, and the flies were abundant. As always, Slinky waited her turn, only staking out a place to do her business once the older dogs were done. After all three had finished squatting and peeing up the wall of the house, I tiptoed across the shit-smeared rocks to palm their leavings with a recycled newspaper bag, dropping my collection into the big green bin provided.

The backyard was enormous, especially by Bay Area standards. A wide patio gave way to the grassy, gently sloping yard, which stretched all the way around the house. It was partitioned from the front of the house by an eight-foot chain-link fence. Not the most beautiful means of containing the dogs, but effective. Ramps, runs, jumps, hoops, and various other tools of the agility-training trade were strategically placed throughout the yard. My job was simply to kick the ball with them for thirty minutes or so, getting them as lathered and well-exercised as I could manage in the short time frame.

They raced for the ball at breakneck speeds, seemingly unconcerned with the location or safety of the other dogs, so long as they weren't threatening to arrive at the ball first. A fierce game of chase followed, with the winner being followed closely up, down, and around the yard until I could convince the ball-holder to drop it at my feet. It took a while to persuade the ball away, and, once I did, I had to be ready to kick it fast and far. They'd charge me if I waited too long. Exercising them was just as much agility and speed training for me.

I've never passed as an athlete, though I gave many a sport my best try. Given my height, most people assumed that I was a basketball or volleyball player. My sister was accomplished at the former, my mother the latter, but I never tried either. Instead, there was the field hockey camp I attended with my childhood best friend; two years on the high school cross-country team, for which I was consistently the slowest runner; and soccer, which was always my very favorite, and which, of all the sports, I was obviously and painfully the very worst at. My parents joked that I spent most of my time on the YMCA teams chatting up my mark on the other team and apologizing to her when I got possession of the ball.

I wasn't much of a runner, which, combined with my height, made me an obvious choice for the position of goalie. But try as I might, I couldn't help ducking and covering when the ball came my way. I was notoriously benched for the duration of the junior-varsity season because I grabbed the ball out of the air while playing halfback. No amount of polite (then increasingly desperate) requests, followed by outright begging, would change the coach's mind in this matter.

This was one indication of many in my developmental years that I neither was physically gifted in sports nor had a competitive bone in my body. It didn't help matters that my best friend was a natural star, recruited to the varsity team almost immediately. Her easy success served to highlight my incompetence all the more glaringly.

Considering that, along with my unconquerable fear of the ball, loose grasp of the rules, and lack of a powerful kick or any approximation of aim, running speed, or ability to challenge my opponent convincingly, it was with a heavy heart that I quit soccer after my failed freshman year. But I never stopped loving the game, watching professional matches on the television and my friend's games from the sidelines instead of playing.

I was glad these dogs couldn't mock my unimpressive handling of the ball. As long as I was moving it around the yard, and quickly, they were happy.

Sometimes, as they tore around after the ball and each other, they'd come straight for me, dodging only at the last minute as though I were one of their training tools, a post to be sliced past as precisely and closely as possible. Many a time, I involuntarily yelped in fear for my vulnerable knees, not knowing if I dodged at the last moment whether that would actually put me in the path of an also-dodging dog, causing us to collide. I stood firm, trusting that they knew their own velocity.

I worried for Slinky, too. In play, she was uncharacteristically willing to give chase right alongside the other dogs, going for the ball just as enthusiastically as they. She was just so small and unpracticed next to her larger and more aggressive cohorts. In the thirty minutes or so we spent playing in the backyard, I tried my best to devise ways that Zipper or Rascal would get to the ball first. But Slinky was young and fast, and those times that she won the ball first, my heart stopped as the others crashed into her with all their weight, trying to wrest the ball away.

In one of my failed bids to aim the prize at the big dogs, I sent the ball sailing over the back fence, striking the neighboring house with a hollow *thwack*. The dogs looked expectantly at the fence, as though the disappeared ball might come sailing back over from the other side. I, too, hoped the solution might be that easy, but no such luck. I had twenty minutes left with them, and nothing to kick or throw.

The fastest way to get the ball back was to scale the fence and throw the ball back over from the adjoining yard. My only other alternative was to walk all the way around the city block and try to eyeball the likely house from the front—assuming I could figure out the right street. These were early days, before smartphones and Google Maps could show me via satellite data where in the world I stood in relation to that ball. I only had my middling sense of direction to rely upon.

I didn't relish the notion of going door-to-door and explaining that a big purple kickball may or may not have landed in their backyard, and could I have it back? The likelihood that anyone would be home in the middle of the afternoon on a weekday was slim to none anyway, and I wanted to limit the number of backyards I was snooping through.

In my career as an animal nanny, I'd have plenty of practice trespassing. It's something no one really tells you; I suppose I had to figure this lesson out for myself. Had I done the math, factoring in how many clients' pets I visit over the number of daily, weekly, and monthly appointments, I'd have known a lot sooner that accidents like a lost ball (or dog or house key) are bound to happen. It's unavoidable that things will eventually go awry. Things that can be righted by stealth and a high threshold for unorthodox behavior.

My first mishap of this sort was relatively easy to solve. During an overnight visit, I was taking a grocery bag full of soiled cat litter from the house to the trash cans in the garage, and the kitchen door locked automatically behind me. Lucky for me, there was a pet door that led into the office. A door big enough to accommodate the Bernese mountain dog with whom I was spending the night. Painfully, cursing my wide hips, I was able to squeeze through the opening headfirst, my successful entry into the darkened room rewarded with a pile of cat barf that I army-crawled through.

In another instance, I left my lanyard of clients' keys on the kitchen table while hooking the dogs up to their leashes. And there the keys remained when we exited the house, the front door locking solidly in our wake. A fence separated the client's front patio from the back deck. Luckily, the house was set down a steep hill, concealing my suspicious behavior from any neighbors who happened to be peering out their front window. I'd tied the dogs to the staircase leading up to street level while I hoisted myself up and over the fence, dropping first onto a wooden railing, and then down onto the wraparound porch. I willfully ignored the sheer drop-off into the boulder-strewn shrubbery beneath the deck, entering the unlocked back door with a pounding heart.

Perhaps most memorably was the time I got stuck under a garden gate. My charge, an overwound terrier, had slid past me as I'd entered the house and proceeded to leap through a hole in the orange construction netting encircling her owner's front garden. The speed with which she executed this maneuver made it seem like she'd been planning her escape for hours beforehand. Maybe she had been. After calling for her from the front yard and then the street, I spotted her in the next-door neighbor's backyard. She was lithe enough to wriggle through the wrought iron bars of the formidable enclosure. I was not. Each fence post was topped with a pointy finial that served double-duty as decoration and intruder-deterrent. I was certainly deterred. I knocked on the neighbor's front door, prepared to admit, hat in hand, that I needed to retrieve a dog from the back lawn. Alas, no one was home. But my vantage point from the front stoop did allow me to spy with my little eye the rather wide gap between paving stone and bottom rung of the gate just adjacent to the house.

It started well. Having removed my fanny pack, I scooted myself faceup and feetfirst underneath the iron gate. And then, in the vicinity of my upper thighs, I got very stuck. My slightly larger-than-average ass was firmly lodged betwixt ground and gate. I felt like Peter Rabbit, dreading the appearance of Mr. McGregor.

I stupidly kept trying to push myself through the opening before I conceded that that was folly, and I then set about unsticking myself in reverse. At the height of my despair, the terrier herself climbed beneath the gate and licked my face. I considered grabbing her and holding her tightly to me so she couldn't get away again, but then I would be without arms to aid in my own escape. I prayed fervently that I could get myself out and that I'd be long gone before the homeowner returned to find me prostrate on her garden path. And, too, that I could recapture the dog without involving the owner or any of my fellow dog walkers. The shame of calling on your colleague to get you out of a stupid and avoidable mishap has little equal.

Painfully, and with great effort, I was able to reverse my way out from under the garden gate, but I still had to call for assistance in recapturing the dog. I was humiliated to admit to my colleague that I had no treats in my fanny pack, and so nothing with which to tempt the terrier. She came to my rescue, her own fanny pack loaded with treats of all shapes and flavors and sizes, all delicious enough to lure the dog back into my arms. This rookie error on my part added insult to the injury of getting trapped under the fence, and never again did I set off for work without an array of dog biscuits on my person.

In these and other unfortunate incidents, I unwaveringly concluded that, as long as there were no witnesses and no one—especially not the owners—found out, only the dogs and I were any the wiser. And the dogs certainly weren't talking.

New as I was to the industry at this moment in time, with three dogs awaiting the reappearance of a ball that they might continue to vigorously chase, and a mere two months into my fledgling small-business ownership, I had no knowledge of these future exploits. I was as yet naive to the necessity of a little B&E, a touch of trespassing, to keep things moving along smoothly.

There was a crossbeam about halfway up the fence, and thanks to my long legs, I was able to get a foot up on it. With an ungainly lurch, I hooked my fingertips over the top of the raw pine planks. I hauled myself up, noting, once I'd gotten both legs over the top of the fence, that there was no crossbeam on the other side. I hung there, doubled over, watching the dogs watching me while my feet scrabbled for any kind of purchase above the sheer drop behind me.

As I swiveled my head around to survey the scene below, I got that distinct sixth-sense feeling that the dogs weren't the only ones watching me. Glancing over my shoulder, I saw someone standing in the picture window of the house that backed up to this portion of the fence. He was looking directly at me, arms crossed as he assessed my precarious situation. And laughing. I held onto the top edge of the fence and pushed off with my feet, crashing through a few young limbs of a spruce tree.

Dusting myself off hastily, I pointed to the ball resting at the foot of the tree.

"Ball!" I mouthed, pointing emphatically. The neighbor, probably in his late twenties and still laughing, could likely hear me through the window.

"This is my ball," I said, loudly this time, holding it up as evidence.

Taking the slobbery ball in both hands, I turned and hurled it over the fence to the waiting dogs on the other side. Not looking back to see if he was still watching, I tackled the fence like Mowgli climbing the coconut tree in *The Jungle Book*. After a few failed attempts, I figured my best bet would be to use the tree as a ladder of sorts. I shimmied up the tree high enough that I could reach out for the fence, first with one armpit braced over the top, then a leg slung up, allowing me to painfully hoist myself up and away from the tree. I was lying atop the narrow span as if it were a surfboard. I pushed myself up such that I was straddling the fence like a horse, ignoring the sharp pain in my groin.

Before pivoting around to descend the far side into safety, I chanced a look at the window, and my heart fell. He was still watching. He gave me a thumbs-up. I dropped out of sight. In front of the dogs, I shook my pants vigorously to dislodge the pine needles that had accumulated there, and then kicked the ball viciously. Next time I'd go the long way around.

Once the dogs (and I, too) were worn out from the running and leaping and chasing and dodging, they took a big messy drink from their kitchen water bowls, which I then refilled and mopped around with a wad of paper towels.

Trying to reward the dogs was an exercise in bravery. I'd learned early on to give Rascal and Zipper their snacks at the same time, tossing them as precisely and simultaneously as I could at their spring-loaded jaws. The scuffle of teeth snapping and treat stealing that ensued when I tried to feed them one at a time usually meant one dog won all the snacks and the other got none. And I'd risk losing a hand in the process. In the brief moments that they were occupied with scarfing their rewards, I gently presented Slinky with hers, which she wisely gobbled with great efficiency. I wondered if, by the end of her training and corresponding puppyhood, she'd be as cutthroat as the others.

Weeks later, the inevitable finally occurred. I arrived to find Slinky crated as usual, but her right front leg was wrapped, and she wore a cone to keep her from worrying the bandage. She'd been railroaded by one of the big dogs during play—thankfully not on my watch, at least so far as I knew. The owners left a note that she'd have to be separated from Rascal and Zipper until she was fully healed. I'd split my time there evenly between backyard play with the big kids, and a gentle and gradually lengthening neighborhood walk with her.

The autumn of Slinky's separation from the others was a wet one. Getting the older dogs cleaned up before they came bounding inside was a near-impossible trick. Standing outside the back door with a towel, we all got soaked by the rain while I tried to get the worst of the mud off their paws. There was no way to keep both dogs quarantined in the mudroom for a cleaning—the three of us wouldn't all fit in that space anyway. Rain-soaked was better than a mud-splattered house, I figured. If I could manage to get one dog's paws clean enough, it had to go right inside, lest it decide to go muddy itself again in the backyard while I cleaned the other. But letting only one dog in the house meant that I had to push him or her through the narrowly cracked door while barricading the other from entering with my body. Trying to get us all in the house in a semidry and clean state evoked for me the riddle about trying to cross the river with a wolf, a goat, and a cabbage.

This precarious scenario was bound to devolve into disaster on one of those wet pre-winter days, and disaster did indeed strike. No doubt anticipating their imminent treats, both dogs went crashing past me into the house, completely filthy with mud up the lengths of their legs and drenching the underfur of their bellies. They'd decided to take a tour through the bedroom, past Slinky's crate, up onto the unmade bed and across the sheets, and back down and into the living room, wagging their tails proudly. Luckily, the office door was closed, and the bathroom was barricaded by a baby gate to keep the dogs away from the cat's litter box. If there was a silver lining to this quick and dirty deed, it had to be the relatively limited scope of the muddy mess.

Slinky had not yet had her walk, and the bed was destroyed, smeared with dirty paw prints and wet dog hair. Before I tackled that, I cleaned the dogs, grumbling the whole time. At least the owners had the forethought to leave out a pile of old towels for this purpose. Generally, I supplied the towels for wiping down rain-soaked dogs, and I was always grateful to owners who offered their own for my use. It often seemed like there weren't enough towels in the world for a day's worth of wet, muddy dogs. The resulting laundry piled up all too quickly, stinking up the car with a fug of damp, dirty dog hair.

I spent the next half hour wiping down the bedding with soapy paper towels, as I had foolishly used up all the provided cloth towels on the dogs. By the time I'd removed all visible traces of dirty dog, the bed was pretty wet. After some thinking and digging, I found a Conair blow-dryer under the bathroom sink and set about drying the sheets and pillows. Slinky watched the whole spectacle, patient as a saint. Once the bed was mostly dry, I tried to affect the disarray of the blanket to look casually unmade, as I thought it had looked before. I felt reasonably sure the owners would never know the difference.

I attacked the muddy trail of paw prints with some multipurpose cleaner from the kitchen and more paper towels, beginning at the backdoor and scuttling along the meandering trajectory through the rest of the house.

Only then could I release Slinky from her crate to get her fair shake at a walk and her chance to get wet and dirty herself. The visit was running outrageously long and would undoubtedly throw the rest of my walks off schedule, never mind the extreme traffic delays that always came along with a little rainfall in the Bay Area. There was nothing to be done for it except to keep moving forward as quickly as possible.

Walking Slinky through the neighborhood was by far the easiest part of the visit, even if it was raining. Our greatest challenge proved avoiding broken glass on the sidewalks of the neighborhood. Her leg was healing quickly, and she pranced along without any indication of a limp, sidestepping the bigger puddles gracefully.

When we finished, I took my time with her on the front stoop, using one of my own towels from the car to dry her off and clean the street grit out of her paw pads. She tolerated me burying my face in the downy fluff behind her ears, still so soft even though it was damp. Zipper could hear us out on the porch, and she scratched at the front door impatiently. But if I took Slinky inside right away, she'd be quickly overwhelmed by the bigger dogs angling for attention. We had to enjoy our one-on-one time apart from the others.

I commanded Zipper and Rascal to wait while I gave Slinky her treat—this one a dietary supplement with ample vitamins for her healing foreleg. We both had to be on point in order to execute the transaction without any thievery of said snack. As I shut Slinky in her crate, I slipped her one extra treat on the sly for being such a good girl.

I had a bad and very embarrassing habit of making up songs about my dogs. I had no reason to feel self-conscious singing to them or making silly professions of love because—unlike my previous bird charges—the dogs weren't talking to anyone. And no one else was around. I often didn't even realize when I was singing to them, so second nature had my one-way conversations with the animals become.

Slinky's little ditty was to the tune of Robert De Niro's song for Jinx, the cat in *Meet the Parents*: "Slinky dog, Slinky dog, I love you, yes IIIIII dooooo."

As was increasingly my custom, I sang to her in farewell. I didn't have a special song for Rascal or Zipper, so they were subjected to Slinky's serenade as I took my leave.

It wasn't until the next week that I saw the webcam, its telltale red light glowing from atop the dresser, trained on Slinky's crate and the room beyond. This was an early model, the Ping-Pong-ball-sized camera set on flexible rubber legs. I couldn't say whether it had been there all along, or whether this was a new addition. The owners' had left no note, and I genuinely could not remember ever having seen it there before.

Had they witnessed the whole bed-blow-drying debacle? I certainly hadn't mentioned it to them, nor they to me. And how much of that would have been visible within the frame of the camera? And if it was new, had they installed it because their bed was weird when they got home last week? Or had they noticed an entire roll of paper towels wadded up in the trash? Was this a live feed, or were they also recording, like CCTV? Could they hear me? How many times had they witnessed me singing that ridiculous song to Slinky?

In a panic over what her owners may have seen or overheard, I wondered if they were watching me that very moment. I smiled and waved at the little robotic ball on its flexible rubber legs, hoping I looked completely at ease on candid camera.

I'd started keeping a running log of insider secrets, a sort of, "Things I wish I'd known then . . ." list of tips. It had started with, "BYOB: Bring your own bedding. And coffee!" This was followed by "Birds can be assholes," and, "No matter how ready you think you are, reread those instructions." More recently I'd added, "Don't go over the fence; go around it." (So many months later, I should've known better than to try going *under* the fence. Same principle, with an even more humiliating outcome.)

And at the end of the list: "Even if the owners aren't home, they are always watching."

Hi Susan,

Has Maddie been getting into the cat food? When I arrived today there was some mystery kibble on the kitchen floor next to her water bowl, and her runny stool suggests she ate something she shouldn't have. I cleaned up the food and put the bag of puppy chow out of reach, but I know she's a smart girl and has her ways of finding forbidden things to eat . . .

Lindsey

CHAPTER FOUR

Wolf Pack

....

I unlocked the wrought iron outer door and then the second wooden door, leaning my weight against it until it gave a groan and opened. A swirling cloud of fine white feathers greeted me at the threshold.

"Maddie?" I called, confused by what or who might have caused the explosion of down that I was now wading through. I knew Susan's son Trevor was not there, given the absence of his ancient Datsun from the gravel parking pad out front. According to plan, Maddie should've been crated in the back corner of the kitchen when he left.

In response to my voice, Ash, the new puppy of the house, came charging into the living room. He slid into my feet, *Risky Business*–style, setting off another flurry of feathers.

"What are you doing in here?" I reached down and rubbed his tummy. Ash was supposed to be sequestered in the backyard, not loose inside the house.

Ash and Maddie were cousins. Or half siblings. Or just siblings. Though Maddie was full-grown and Ash just a pup, both were descended from the wolf mother and German shepherd father who lived in a chain-link partition behind the house across the street, along with an ever-changing assortment of their offspring.

Ash was all white, like his dad. Maddie took more after her mother, the wolf unmistakable in her face and frame and coloring. The wolf mother had a permanent sneer from an old bullet wound—whether it was inflicted there in Oakland or before she was domesticated, I didn't know. It gave her a sinister look, in contrast to Maddie's sweet face and gentle, intelligent eyes.

For all of her good-natured playfulness and irresistible lovability, Maddie was not an easy charge by any stretch. The wolf in her introduced all kinds of complications not faced by most other dogs. At least not in such an extreme combination of characteristics. As was common in wolf dogs, she was way too smart for her own good, strong as an ox, and a superior escape artist. She also had a highly sensitive stomach. Her digestive issues were not necessarily endemic to this hybrid but were further compounded by the specific and often divergent dietary needs of a half-wolf, half-canine. All of these challenges were intensified by the recent introduction of Ash, also a wolf dog, into the household.

"All right mess-maker, did you get into a pillow fight?"

Ash padded behind me into the kitchen, the feather storm making us both sneeze. I could only assume that he got ahold of a bolster or a blanket while he was on the loose that morning, and then did what any self-respecting wolf pup might: destroy. Whatever it was he got into, the feather-fall grew thicker the farther into the kitchen we ventured.

In a small dining nook off the back of the kitchen, Maddie was indeed crated. She looked like a canine version of the Abominable

Snowman, her mottled dove-gray and cream coat further lightened by a head-to-toe dusting of fine white down. Looking adorably innocent, Maddie dipped her head again to tear at the remains of what was, until recently, a mattress pad or comforter.

"Oh, you're the mess-maker!" She nuzzled at my fingers through the bars of her crate, licking them with a feather-flecked tongue.

"Did your silly mama give you a feather bed? Those don't really mix too well with wolves, huh?"

I was quickly trying to calculate the next best step in this mess. Every time Ash moved, he sent another plume of feathers floating even farther afield. I could put him back in the yard and leave Maddie crated while I cleaned up, but I really needed Maddie out of her crate and out of the way so I could bag the remains of the cushion. She and Ash weren't really supposed to be outside together unsupervised, though.

"All right, kiddos," I said, resolved to leave the cleaning until later. I unlatched the door of Maddie's crate, and she tore down the hall to the flight of stairs that led to the bottom floor and backyard. I knew proper procedure would have been for me to make them both sit and stay at the top of the stairs and wait there for me to give them permission to proceed. Sometimes you have to pick your battles.

At the base of the stairs that led to the backyard was Trevor's room. His door was always shut, leaving the small landing in gloom. Today, however, the door was wide open, and I thought for a moment that Trevor was perhaps home after all.

"Hello?" I ventured, suddenly feeling self-conscious about talking to the dogs so freely. I mentally reviewed what I'd been saying, hoping none of it was too ridiculous. Or openly critical of Trevor or Susan. The dogs were clamoring at the back door and wouldn't be ignored.

"Okay, okay, out you go," I said, sneaking a glance over my

shoulder into the dimly lit room as I unlocked the back door and released the hounds.

His room was spartan, almost alarmingly so. Mattress on the floor, thin blanket thrown aside, guitar leaning against the wall, plain curtains closed. And no Trevor in sight. I relaxed a little.

Interesting that their dog got the feather bed and he had a bare mattress on the floor. Maybe his squatter aesthetic was by choice.

His open door had distracted me from noticing the window to the right of the back door, and Ash's only point of entry from the yard into the house. The window was broken—and had been for a while—so Susan or Trevor had propped two brooms in an X over the opening. Ash clearly figured out that he could knock these aside in order to climb through the window. I scoffed under my breath at the notion that broomsticks were sincerely intended to keep a wolf pup outside.

This client—Susan and her son—were not clients of my own, but one of the many jobs I took on in a subcontracting capacity. Thus, I didn't have a direct dialogue with them but deferred to the authority of the colleagues I was working for. I know, however, due to my coworkers and my near-constant communication about these dogs' well-being, that they'd requested many times that Susan fix the window. Word was that Trevor was supposed to take care of it.

I peeked my head through the window to check on the dogs and saw that there was also a chair adjacent to the open window, further enabling Ash's access from the patio. My deep and heartfelt eye-rolling was interrupted by Ash starting to take the trademark squat. I caught on just as Maddie did, and I could see her anticipating the treats he was about to drop for her to gobble up.

Even in my short career of dog walking and animal nannying, I'd already encountered many dogs with strange and sometimes dangerous eating habits. There was the American bulldog who ate his owners' gym socks at every opportunity. Or the corgi who preferred his dry food with a helping of raw broccoli and cauliflower on top. I pet-sat one weekend for a Rottweiler that I dubbed the Mulcher. She had a penchant for eating the dried leaves that littered the back patio, where she spent her days. Then she'd come inside and leave a loose, detritus-filled dump on the carpet. This I cleaned up with the industrial-grade Hoover the owners helpfully left out for this very reason. In all of these instances, the owners warned me in advance of their pets' unusual predilections, which helped enormously in both preventing a problem but also in understanding what the hell I was looking at when I showed up to take care of said pet and clean up his or her messes.

In the instance of Penny, a newly adopted Lab-terrier mix puppy, I hadn't gotten the benefit of full disclosure. I was running blind when, on one of our neighborhood walks, Penny started to drag her butt on the grass. Worms? I didn't think so. I knew from her paperwork she'd been spayed and vaccinated and microchipped prior to adoption. Along with all of those procedures, deworming was standard. Yet another detail I'd gathered during my glory days at the pet store, since I was the one who administered the thick yellow tonic to the incoming animals, and then got to flush the worms down the industrial disposal when they came out the other end.

Farther along on our walk, Penny started dragging again, and I noticed that something bright red was crowning beneath her perky tail. Blood? Intestines? I was freaking out. I got down on my hands and knees for a better look, as more and more red bloomed from her bottom.

Upon closer inspection, I could see that this was something inorganic. It had a little white tag on it, the tiny lettering hard to make out. I then did what one should never, ever do in this situation. Chalk it up to inexperience, but I bagged my hand, and I pulled.

Lucky for me, the object had not become entangled in Penny's innards, and it came out cleanly (a relative term) in my hand.

JLo Intimates, the tag read. I was holding the owner's undigested thong in my hand.

I knew from my days at the pet store that there's a technical term for Maddie's own dietary anomaly. We had many an appalled dog-owner coming into the store to complain about this unattractive habit some dogs have of eating shit, a condition called coprophagia. We'd direct the customer to aisle two, top shelf, where they'd find tablets for this affliction. I can't say how helpful the pills were in deterring their dogs, but they certainly didn't work on our darling wolf mix. Bless her, this disgusting habit absolutely tore her already-sensitive stomach apart. Whether it was her own, or anyone else's, she was indiscriminate in her partiality to poop. And if I didn't move quickly, she was about to make a snack of Ash's.

"No!" I shouted, leaping through the open door to place myself between Ash and Maddie. I didn't have time to grab the shovel in my mad dash, so I collared Maddie and took her with me to collect the pooper-scooper. But first, I sat down on the patio chair and looked into her precious face.

"Hey, you know better! No, ma'am! No poop for you." It was extremely hard to resist kissing her soft muzzle, but this was not a time for positive reinforcement. I was being as stern with her as I could manage, resisting her charms with all my might. She was a love, but this behavior had to stop.

Maddie was on a very strict diet of boiled chicken and plain white rice, a combination that seemed to meet the three-prong requirements of going gentle on her stomach, providing protein for her wolf half, and satisfying the domesticated dog in her that could process carbs with alacrity. If she weren't possessed of such a tender tummy, I feel sure she could have taken down all kinds of meat. I'd worked with a woman who regularly fed her German shepherd raw steaks and swore by the benefits for the dog, if not the expense, of such a high-quality regimen.

This rice-and-chicken diet, while mostly successful in keeping Maddie fed and comfortable, her stools healthy and firm, was all-too-often interrupted. If not by the shit-eating she engaged in, then when, for example, she figured a way to get into Ash's delectable puppy kibble. Or the food of Susan's many cats, whose dishes sat out on the front patio, tempting Maddie every time she passed by.

I kept a firm grip on Maddie's collar, and we returned to Ash's pile. She was way too fast for me, and there was no way I could let her go and clean up after him before she beat me to it. I quickly scooped up the offending mess and only released her collar once I'd dumped it into the trash can left out on the patio for just that purpose.

Susan tried her very best. I knew that. She loved Maddie, and now Ash, fiercely. She closely followed the recommendations given by myself and the colleagues with whom I shared Maddie and Ash's care. Except when she couldn't. Susan worked irregular shifts and long hours, and she relied on Trevor to manage the house and the dogs in her frequent absences. This meant the fifty-pound bag of Ash's food was often left out and open in the kitchen, where Maddie could dip her face right into it like a horse's feed bag. Or else Ash was fed in plain view of an uncrated Maddie, setting him up to be body slammed aside by his much larger and hunger-motivated housemate.

What seemed at first like carelessness or ineptitude on Trevor's part was looking more and more like sabotage—of Susan's intentions, us dog walkers' efforts, and Maddie's overall health and wellness.

All I really knew of him was that he went to community college, which explained his unpredictable schedule. He rarely spoke to me, and he made eye contact even less than that. When I would make an effort at conversation, or ask him questions about the dogs, I rarely got more than a mumbled reply.

The backyard was basically a grassless wasteland of dust and rocks, enclosed by a ten-foot fence. In the corner, there was a stand of bamboo that Maddie loved to hide in. She'd carved out tunnels in the thicket that we couldn't get to, and she would go to the bathroom there, where we couldn't swoop and scoop right behind her. It was very frustrating. Try as we might, my fellow dog walkers and I had not yet succeeded in getting Susan to chop it all down. Susan had agreed that it needed to happen, of course. On multiple occasions. She said Trevor would do it.

And so there we were, bamboo forest intact.

I threw a filthy, fuzzless tennis ball for the dogs a few times to get some of their energy out and refreshed their water from the hose coiled at the side of the house. They each took a long drink before we headed out on our walk. Both dogs were enthusiastic pullers, still learning how to heel on command—though with enough practice, they'd eventually do it without me even having to ask. They wore pronged collars to which I attached their heavy-duty canvas leashes.

Ash was still at prime learning age, while Maddie should have mastered the simple command long ago. My hunch, based on her extreme intelligence, was that she fully understood; she just didn't care to comply. Maddie was, more than any other dog I cared for, that most devilish combination of cute and headstrong.

They went on-leash by the back door. Then, we practiced heeling at the base of the stairs, again at the landing where the stairs turned, and at the top of the flight. It didn't go well, both dogs sneezing heavily at the feathers drifting down the stairs like snow, and otherwise generally disregarding my instructions in their enthusiasm to hit the road.

At the front door, I had to be extra focused and firm with the dogs. The walkway leading up to the street was lined with cat dishes, scattered in and among the broken flower pots, a rotting bench, cobwebbed watering cans, and a dusty hose, its nozzle sitting in a dirty bowl of water left out for the cats.

I knew Maddie—and Ash, too, for that matter—would love nothing more than to clear each and every plate, tantalizingly crusted with kibble and dried Fancy Feast or some other cheap brand of wet food. Poor thing, Maddie was so desperate for anything other than boiled chicken and rice. Like a hungry girl on Weight Watchers, she had very keen radar for the presence and location of any and every morsel she was not allowed to have.

Keeping both dogs tightly reined in, one on each side of me, we made it up to street level without incident, assorted cats fleeing in our path. Dave, Maddie and Ash's former owner and current caregiver to their parents, lurked at his mailbox.

"Hey, girl!" he called. "When are you gonna come walk my dogs?" He'd been asking me this for weeks, and I hadn't yet mastered an effective deferral. I tried not to look uncomfortable.

"Pretty busy, Dave. Sorry." I tried to soften my answer with a laugh, which came out sounding like a strangled cough.

I didn't consider myself above anyone's business; I wasn't so flush as to turn away clients. In almost every case, if they'd pay, I'd walk. Or spend the night, or give the dog or cat a bath, or take them in for their vaccinations. I was indiscriminate in my willingness to earn money helping in whatever way I could for whomever needed it.

This went somewhat against the unspoken credo within the professional sphere of animal caregivers. Based on my training at the hands of my mentors, the first (always free) meeting with a potential customer was just as much me interviewing them as the opposite. I absolutely understood why this had to be so. Entering their house often and alone to care for one or some of their most beloved companions, countless things can—and do—go wrong, even under the best circumstances. Good communication and mutual trust are absolutely essential and are a baseline requirement for a functional relationship between pet-care provider and pet owner. This was Dog Walking 101. Also, it was far preferable if you and the animal's owner were on the same page about what was best for their pet's health and happiness. And while it seems like this would always be the case with any reasonable person who loved—or even liked—their pet, such easy agreement was not always a given.

While it was obviously better to be extra discerning about my clientele, work was work, and I was having an increasingly hard time turning anything away. My business model of in-home pet-sitting and individual neighborhood walks, versus the five-dogs-at-a-time (at five times the take-home pay) group walks, was the most time-consuming and least lucrative approach in the industry. Plus, I was getting paid the reduced subcontractor rate for the majority of my appointments. Subcontracting was great because I was taking care of a client list that I didn't have to build or cultivate myself. But it meant a lot less money.

This was yet another rookie error, building such a flawed business model. The upshot was that I needed to pull a much larger income than I was. Thus, I was more willing than most to overlook some of the potentially annoying, worrying, or irresponsible aspects of a paying patron. Even considering my hyper-permissiveness when it came to high-maintenance or otherwise unappealing clients, I could recognize that Dave took all of these yellow-flag-flying attributes to a new level. He was solidly red flag in my book.

Just last week, the German shepherd dad was savagely attacked by one of his adolescent pups, presumably in an effort to usurp the dominance of the paterfamilias. According to Susan, it did not end well for the dad. Dog fights aside, the wolf mama was constantly pregnant with a new litter, and Dave's driveway was always host to a revolving cast of cars. Whether their owners were there about dogs or drugs or something else altogether, I didn't know. And I didn't care to. The relationship between the neighbors and Dave was tense at the best of times, no one owning up to the frequent calls to animal control and the SPCA about the escapes by (and the growing number of) the wolf dogs, the conditions the dogs lived in, the constant barking and howling, and so on and so on. Every time a siren passed within ear shot (which, in this East Oakland neighborhood with its rate of crime and proximity to the freeway, was frequent enough), the family howled in concert, their collective call astonishingly loud and primal.

Susan certainly didn't adopt Ash as a matter of convenience or ease, but because she couldn't stand to watch the wolf/canine/human drama unfolding across the street without doing something, however small. Ash was like a refugee, spared the horrors of his family situation by the savior Susan. I respected the hell out of her for intervening in whatever way she could. Even if it was abundantly clear that these wolf dogs were more than she could currently manage.

That's why she hired dog walkers. It was our job to lessen her load and make the dogs' lives more comfortable. If only we could figure out how to peacefully, respectfully manage the loose cannon that was Trevor.

I exaggerated the effort it was taking me to manage Maddie and Ash in order to avoid engaging in further conversation with

Dave. Around the bend and out of his line of sight, I unclenched a bit and let the leashes out to allow the dogs a sniff and a pee. As the male in the group, Ash was naturally more inclined to mark every telephone pole, pinecone, and piece of garbage.

We practiced heeling some more, to little effect. Part of the problem was the lack of incentive for Maddie. Short of carrying around chunks of boiled chicken in my pocket, I didn't really have any tummy-friendly treats to offer her. My verbal praise, however effusive, could hardly be considered reward enough to change her behavior of pulling at the leash. Without a snack to offer Maddie for good behavior and well-executed heels, I could hardly give Ash a snack right in front of her. That would be too mean.

Despite the uneven economic status of its residents, the neighborhood we walked through made for a very pleasant thirty- to forty-minute walk with the dogs. It was hilly, which gave the dogs—and me—a good workout; heavily wooded with evergreens, blocking out the sounds of the nearby highway; and virtually traffic free in the middle of the day, most of the residents either out at work or at home on the couch watching TV.

We walked the usual two-mile loop, unimpeded by the large intersections or construction sites or broken glass and other debris that littered some of my other routes, enjoying instead the dappled winter sunlight that filtered through the soaring trees.

The streets were laid out like concentric circles, turning back around on themselves in ever-widening loops until the outermost ring abutted the access road that led eventually to the on-ramp. The humble single-story homes were mostly set back from the street, many of them behind fences or gates, given the more-than-occasional crime in the area. Despite the bucolic quiet during daylight hours, these streets were no strangers to after-hours activity. Proximity to the highway was no doubt a contributing factor, providing an easy getaway by car. Thankfully, theft and vandalism remained the primary problem, unlike some of the other parts of town I frequented, where actual assault, occasionally armed, was the bigger concern.

No matter the neighborhood, I was always happier to perform my dog-walking duties during daylight hours. Here especially, on Maddie and Ash's turf, where the relatively peaceful and rubbish-free thoroughfares lent a tranquil tone to our walks and banished the notion of any immediate threat. So long as Dave wasn't lingering, asking questions, and watching me walk.

As it happened, I spent a good portion of each week beating this particular path around this neighborhood. In addition to Ash and Maddie, the colleagues I contracted with had two other local clients that I visited frequently. Given that I repeated this route so many times, day after day, I was especially grateful that it was comparatively safe and easy to navigate.

Down one side street lived Ralph, a jaunty little corgi, and his elderly owner Edie, whom I loved like my own grandmother. Of course, it wasn't professional or appropriate for me to feel this way, much less make that apparent to her. When I came to care for Ralph, I was always happy to stay a little longer, chatting with her from the comfort of the plush rose-colored Barcalounger. Sometimes I helped her fix the remote, or retrieve something that had rolled beneath the couch. If she offered a cup of tea, I always accepted, time allowing.

Further along Maddie's street lived Gabby—a terrier mix who loved barking perhaps more than anything else—and her geriatric companion, Stuart. He was a slow-moving, fluffy black mutt of indeterminate provenance who happily went wherever Gabby did, but ten feet or so behind. It was Gabby who escaped on that unfortunate afternoon when I became lodged beneath the neighbor's garden gate, a memory that returned every time I walked by the scene of the crime.

Of course, Maddie and Ash have never met Ralph or Gabby or Stuart. Neither had Gabby and Stuart met Ralph, or any other variation therein. Gabby barked her heart out anytime I passed her house with one of my other charges, her dog radar going crazy, so she perhaps was the only one who would be any wiser about my fraternization with other animals in the neighborhood. I always tried to make the dog I was caring for feel like my favorite and sole focus for that thirty minutes or hour. I had no other children, only them. And whomever I happened to be with at that moment, I loved them best of all.

When we got back to the house, I could hear the shower running in the small bathroom. I couldn't see that there was any change to the fine coating of feathers over every surface in the upstairs, and I wondered if Trevor intended to ignore it like I had. I flushed at the thought of him showering on the other side of the thin door, and I realized that it had been too, too long since I'd been with someone in a remotely romantic capacity.

My last relationship predated my move from the Southeast to the West Coast and ended disastrously, but predictably. I'd held out hope for us when I moved; he then met someone else and failed to inform me, and we'd had a messy reunion when I was home for Thanksgiving just a few short weeks before. He brought his new love—unannounced and uninvited—to a party at my sister's house, and we hadn't spoken since.

I had neither the energy nor the interest to seek out a replacement or even a rebound. My heart was still quite sore from the abrupt rejection, and I didn't quite trust my judgment. Not to mention that I didn't have the first idea where I might meet someone.

Since moving to California, my days and nights were filled with dogs and cats and birds and fish. The only evidence I'd had that I was even visible to anyone of the opposite sex came in the form of a comment in the snack aisle of the grocery store, when a fellow customer asked me if I found that short men were always attracted to tall women. Without thinking, I said, "No," looking down at him. "I don't find that."

But then, my experience was unfortunately limited to college-age almost-men, and they'd rarely articulated what, or who, they were attracted to. They left us ladies, short and tall alike, to puzzle that out for ourselves.

Heartsick or not, I was keenly—uncomfortably—aware of how hot and bothered I was at the thought of Trevor in the shower. This was Trevor, for god's sakes. He couldn't even feed his dogs without screwing something up. I couldn't deny that he bore a slight resemblance to a young John Lennon. In fact, he was kind of a dead ringer for Sean Lennon. Not unattractive by a mile.

I busied myself with the dogs, unhooking them each in turn from their collars and leashes and giving both a vigorous butt rub, sending them into wiggling circles of ecstasy and garnering many appreciative kisses on my hands and arms. I tried not to think about what might have been in Maddie's mouth last and was now coating my slobbery wrist.

When Trevor emerged from the bathroom, wrapped at the waist in a white towel, I was elbow deep in the fridge, looking for some Maddie-friendly food. Usually Maddie's chicken and rice were in Tupperware on the top shelf. Or, if it hadn't been prepared for the week, the family-sized flat of raw chicken breasts would be hard to miss. But there was nothing resembling anything remotely edible for her. Just a half-empty tray of what looked like enchiladas and a container of something stinky and unidentifiable. I couldn't tell if it was intentionally green (i.e. guacamole, spinach, seaweed) or had become that way not-so-recently.

I hadn't bargained on having a conversation with Trevor while he was in nothing more than a piece of terry cloth, and was grateful to be facing the fridge as he passed through the kitchen toward the back stairs.

I focused instead on containing the feather mess, figuring we could discuss Maddie's food situation when he was fully clothed. Trying to sweep tiny weightless feathers is like digging a hole in wet sand, though, so I gave up after stuffing the shredded remains of the cushion into a trash bag. What I really needed was a vacuum cleaner for this job.

I listened for any indication that Trevor was coming back up the stairs, but I couldn't hear much over Ash's whining about his empty dish. And then I heard the faint strains of a guitar being strummed.

Trevor wasn't coming back up.

As a last-ditch effort, I checked the stove top for a pot of boiled rice, to no avail. The soup pan looked like it held the day-old remains of oatmeal. I sighed, out of ideas. Only then did I see the note stuck to the fridge door, scrawled in Susan's loopy script: *Ran out of chicken. Trevor will make rice before he leaves. M can have some of A's kibble.*

"But she can't." Ash's puppy food, never mind that it was for puppies, gave her the worst kind of bloody, mucous-striated diarrhea. I'd struggled with IBS for as long as I could remember and empathized with Maddie's poor, sore bottom and churning stomach when she ate the wrong stuff.

I crated Maddie while I gave Ash his snack-bowl of chow, allowing him to eat unmolested. I shut the cabinet holding his food a little too firmly, taking out some of my frustration on the sticky latch. I felt so guilty watching her watch him with cocked ears, eyes alert, as he scarfed down the food.

"I'm so sorry, baby, I don't have anything for you yet. Trevor never made your rice." I glanced at my watch, stressing the minutes that were rapidly running by and making me late for my next walks.

Before thinking it through too clearly, I marched downstairs and knocked on Trevor's now-closed door. The guitar stopped, and I heard shuffling before the door opened a crack.

"Uh, hi, Trevor." The opening widened. He was blessedly back into jeans and a T-shirt.

He looked down and just left of my knees.

"Do y'all have a vacuum cleaner? Sweeping the feathers isn't exactly working."

"Oh, I'll look," he said, so softly I could barely hear him.

I moved aside, assuming Trevor was going to go look for this mythical beast that was the vacuum cleaner, but he made no move to leave his room.

"Your mom left a note about Maddie's rice? I couldn't find any that had been cooked . . ." My voice trailed off, and I cursed the way I was meeting passivity with passivity.

A pause. Maybe he was thinking about it.

"Oh, right," he said in monotone. "I'll do that."

"Okay, cool. Are you getting the chicken for her, or is your mom?"

"Um, I'm not sure."

I was trying not to scream.

"Can you get some more this afternoon? I think she's pretty hungry."

Nothing.

"Okay, thanks. And the vacuum cleaner . . . do you want me to clean up before I go?"

"S'okay. I'll do it," he said, the door already starting to close.

"Okay. Thanks!"

As I turned, the door clicked shut behind me, and I saw the stupid ineffectual broomsticks peeking through the window frame. I hadn't mentioned that to Trevor.

I contemplated conversation part two and decided I—and Maddie and Ash—would be better served if I just mentioned it to Susan directly. All of it. I still hadn't quite gotten used to the binary nature of my clientele. It seemed like the owners were either wound really tight over the conditions in which their pets lived and the specifications for their care, or else they weren't paying nearly enough attention. In either case, when negotiating with helicopter and laissez-faire parents alike, tact and diplomacy were tantamount to a successful relationship between client and service provider. These were skills I was still honing, though not nearly fast or masterfully enough.

Ash was long-since done with his kibble snack, so I rereleased Maddie from her crate to have free reign of the house along with him. I double-checked that anything edible and verboten was out of reach. With Trevor home, they didn't have to be separated, though I had less and less faith that there was much difference between his supervision and none at all.

If I didn't have a packed afternoon of walks scheduled—and if I'd had more than three dollars and change in my checking account—I'd have driven to the store and bought the chicken myself, so little confidence did I have in Trevor's assurances. Or motivation. Or memory. But I was supposed to be up the hill at Edie and Ralph's, and my tardiness had already eliminated any possibility of a cup of tea or even a chat.

Instead, I took Susan's note down from the fridge and flipped it over, summarizing the feather bed's sad demise, adding some high notes from our walk, and trying to tactfully underscore Maddie's hunger, gently reiterating how hard the puppy chow was on her system. I also tacked on, at the very end, a question about the unrepaired window. It felt futile.

I would, of course, also be notifying my colleagues immediately of the day's events. Hopefully I'd be able to steal a few moments during my walk with Ralph to bring them up to speed. There was always strength in numbers, and I knew we were very much on the same side of this battle for the dogs' well-being. I also knew their conversation with Susan would be far more pointed, more forceful, than my polite note. As the primary service providers, they wielded more power and persuasion than I did as the subcontractor. Were this a construction job, they'd be the foremen to my plumber.

I posted my note in the same spot on the fridge, kissed the pups goodbye, and locked both doors behind me. A few feathers escaped the clanging outer door, gusting across the dirty patio toward a mammoth sleeping marmalade cat, his blocky head resting on giant paws.

I took a deep breath, trying to put the events of the last hour out of my mind so that I could focus on the rest of the day and the dogs on my schedule, and not stew about Maddie and the chicken and the brooms and the feathers. At least Dave was nowhere to be seen; I didn't trust myself to be civil at that moment. I could only hope the afternoon ahead had few surprises in store. And that, when I arrived here tomorrow around this time, everything was in far better order for the little wolf pack of two.

To: "Tom", "Patty"

Subject: Daily update

Hi guys,

Thanks for the update on Boo and his bum salve.
I'll be sure to apply it and send along any updates
after my visit. I had another dream about Blondie
and Buster last night: they bit each other's tails off
in a dog version of a scene from *Kill Bill*. I think I'm
losing my mind.

End-of-month invoice to come shortly!

Your (questionably sane) subcontractor,

Lindsey

CHAPTER FIVE

The Farting Greyhound

....

Around the time that I shacked up with Charlie, the overnight gigs were starting to wear on me. I felt like a doggie date. A call girl for the canine contingent. *Pretty Woman* for pooches. But the overnight jobs—aside from being my niche within the local industry—also provided a roof over my head. Unless I radically altered my business model and client base to center around the more lucrative group, off-leash walks, I couldn't afford to give up the slumber parties. Never mind the difficulty of drumming up that many new clients, the greatest obstacle between me and the group-off-leash paradigm was actually me.

I'd heard too many horror stories of colleagues getting fined for running afoul of park rules, or getting sued by a client because the dog bolted into traffic and died. No matter how much voice control a walker claims they have over their dogs, something is bound to happen at some point. I much preferred the control a leash afforded me.

Charlie and his owner, Katherine—some kind of advertising executive, single, late thirties—lived in an ultra-manicured corner of Piedmont. At our introductory meeting, Katherine had said to me, "My house isn't even that nice. I hope you won't mind staying here."

Her single-level home was nestled among the expansive mansions of the Oakland hills: a neighborhood of flawless lawns, long driveways punctuated by luxury cars, high fences, and even higher mortgages. Porta potties dotted the sidewalks, accommodating the countless contractors that swarmed over the estates by day, to perfect what was already arguably perfect.

So by some standards, one might have called her house, without a second floor or five-car garage, humble. But the long hallways of shining hardwoods, the monogrammed towels and sheets, the French-provincial living room and meticulously landscaped terrace garden, and even the dog, all smacked of quality. And a cozy relationship with some well-paid interior designer.

At roughly six foot two, Katherine was one of the few women taller than me that I'd ever met. I was inclined to forgive her feigned modesty about her house and like her for being one of us. "Large and in charge," as my sister says. Katherine surely understood the plight of the outrageously tall woman, and I took note at our meeting how she did it with such class. She was clad in ballet flats, jeans, and a luxuriously soft-looking cropped sweater. Cashmere, probably. Her short chestnut hair was shiny and bobbed, and she had very creamy skin—something I've aspired to and still fall horribly short of. I've always sported more of a well-scrubbed Scottish complexion. Florid, one dermatologist called it.

I thought Katherine and her greyhound made a handsome pair; equally long, lean, and refined. For his part, Charlie had a glossy sand-colored coat, great teeth (another thing I could only dream of), and a spring in his coltish step. He could have stepped out of one of the old English paintings in the dining room.

At that first meeting, Katherine had started the tour in the living room.

"This is Charlie's teddy bear," she said as she gestured toward a slobber-dampened lump on the green tartan dog bed. "Make sure he has it when he is outside during the day, and at night when he is in his crate. He needs his teddy with him."

Katherine glided from the living room, through the formal dining room, and into the kitchen. She opened the freezer to reveal an array of homemade dog treats. There were jerky sticks and dog biscuits hidden in frozen cottage cheese, peanut butter–filled toys, and cheese-stuffed kongs.

"But his favorite," Katherine said as she opened the refrigerator below, "is egg. There are two egg crates here with hard-boiled eggs. He eats them with his morning and evening meals, crumbled over three cups of kibble. Then he can have a treat from the freezer."

Listening to all of this, I couldn't imagine the havoc it wrought on Charlie's digestive system. All that dairy. And on top of the outrageous amount of food he was fed in a day! His craps had to be the size of smart cars. I was no stranger to dogs' unorthodox dietary habits, but usually the questionable food wasn't being pushed quite so forcefully by the owners themselves.

I followed Katherine down the hall to the guest bedroom.

"This is where you will sleep," she said as she gestured to the matching twin-sized sleigh beds, dressed in monogrammed yellow gingham. "Charlie's bed is through here." She continued down an adjacent hall into the master suite, where Charlie was already waiting for us. He was sprawled across Katherine's unmade king, his head on her pillow.

"His crate was too heavy to move, but perhaps he'll be happier sleeping with you in the guest room. You can put his cushion in there if that works." She turned off the light in the bedroom, and I followed her back through the maze of hallways and into the laundry room.

"This is where you'll find his leash, his fleece, doggie bags, and his toothbrush and toothpaste. I brush his teeth once a week—he loves it—so if you'd like, you can certainly do it while I am away. Twice, if you can.

"There's just one other thing. The previous sitter broke the washing machine. Apparently she was doing her laundry while she was here with Charlie and the washer overflowed." Katherine rolled her eyes and waved her hand dismissively, like laundry was something she'd rarely dealt with and definitely wouldn't tolerate of someone living in her home while she was away. I wrote in my notes, *NO LAUNDRY,* triple underlining the words.

"I'm traveling through Italy—Venice, Florence, Rome—and the only way to reach me would be through my travel agent. That information is by the telephone, but I don't expect you'll have any reason to use it. If there's an emergency, just call Charlie's vet. His office and home numbers are by the phone as well." I wrote, *Unreachable,* but mentally I was scribbling, *Don't touch anything! Don't eat any of her food, don't use the towels, don't drink on the furniture, and don't drool on the pillows. Don't screw this up!*

So there I was in the palace of perfection. When I arrived the first night, I brought Charlie in from his private courtyard where he was lounging on the stoop of his custom doghouse. It had a ramp up to the wrap-around porch, eaves, and even contrasting trim paint.

When I gave him his three cups of kibble with egg, he shoved his long snout through his bowl, eating every last big of egg and leaving the rest behind. He walked away with bits of yolk on his face. Judging by the loosely formed piles under the bonsai trees outside, he really did take monster shits.

After dinner, Charlie trotted around the house with his bear fixed in his jaws, making a strangled whining sound. For a while he sprawled across Katherine's bed with his teddy next to him. He didn't show any interest in me until I pulled out his leash from the basket in the laundry room.

I strapped him into his hunter-green fleece, and I had to admit, he looked handsome in his walking outfit. Given my experience with other greyhounds and their finicky reaction to rain and chill, I figured he'd probably appreciate this extra layer. The February evening was crisp, though certainly not as cold as it could get in late winter.

With daylight savings still months away, the street lights were already on at seven o'clock, bathing the hushed street in a tangerine half-glow. The construction workers had gone home. The porta potties stood at intervals along the sidewalk like crooked teeth that needed pulling. Charlie stepped over the threshold willingly enough, onto the mossy cobblestone path that led to the street. At the curb he stopped short, though. He lifted his long snout to the air, nostrils flexing.

"What's up?" I asked as I tugged gently on the leash, already a little chilly and ready to get my blood pumping with a vigorous walk.

This was the very best neighborhood to walk in, during the day or in the dark. It seemed utterly inoculated against anything threatening or unsavory. No garbage, no overgrown grass, no tacky mailboxes or junky looking cars. This bucolic pocket of the city was a far cry from those neighborhoods littered with broken bottles and the occasional needle, used condom, or KFC box full of chicken bones that just begged to be choked on by a sniffy, curious dog. Which all dogs are. There, daylight provided no assurance that bad stuff wasn't going down.

I clicked my tongue at him. "Giddy up, Charlie boy. Come on!" He stood completely still.

All I could think was that all the egg and cheese and kibble must be blocking him up. I'd be constipated, too, if I ate like that. He had to walk it off—get it out—before he went to sleep. But he was clearly not in a walking mood. I engaged him in a brief tug of war, cooing soothingly. I was nearly parallel to the ground, pulling with all my might, before he budged an inch.

I settled for walking him up and down the cobblestone path over and over again, hoping each time when we reached the sidewalk that he'd relent and continue on down the road with me. After about the eleventh lap, I tried another pull-a-thon with him. But as he skidded across the pavement toward my firmly planted feet, I was afraid that I'd hurt his neck or break one of his toenails or scuff his footpads.

Charlie ended up fertilizing the bonsai in the courtyard, and then we sat on the kitchen floor while I tried to get a blob of tree sap out of his velvet fur. I'd already tried using a soapy paper towel and was still rubbing at the sticky fur, trying to scrape the resin off his short, fine hairs. In my determination, I yanked out a pea-sized patch of fur along with the sap. Charlie yelped and pissed a little on my leg before skittering down the hall into Katherine's room.

The next morning, I found Charlie huddled in the back of his crate, his cushion soaked in pee. Not even one day gone and I already had to use the washing machine. Ignoring my all-caps note to myself about not doing laundry, I stuffed the waterlogged cushion into the washer. I added a healthy scoop of liquid detergent and flipped the temperature to hot.

The laundry room flooded in no time. Frothy waves of piss-tainted water belched from the lid of the washer, spilling over onto the floor and making its way closer to lip of the laundry room door toward the hardwood floors beyond.

Clean as her house was, Katherine seemed to live without any cleaning supplies—no mop, broom, rag bag, bucket, or sponge— save two ultra-absorbent rolls of paper towels under the kitchen sink. The paper towels were soft enough to bathe with, and I used every last one as I tried to stop the flood that was now starting to leak down the hallway. I weighed my options: call a travel agent, or a vet?

I'd quit the business before using one of Katherine's fluffy monogrammed towels to such a filthy end. And it, too, would have to be laundered. There was nothing towel-like in my car, just some Starbucks napkins and a fleece hoodie. With the recent reprieve from rain, I'd washed all the rags and towels for wiping down the dogs and never returned them to the trunk.

Back inside the house, I found a faded beach towel in the back of the hallway linen closet. After squeezing and re-squeezing it into the laundry-room sink, I was able to clean up most of the water. Half an hour later, sweaty and a little sudsy, I stood back and surveyed the reasonably—manageably—soggy floor. The hardwood floors, at least, were dry.

I hand washed the beach towel in the deep sink adjacent to the washer and threw it in to dry with Charlie's bed.

Soon after getting into animal nannying, I started having anxiety dreams about work. The common theme of all the nightmares was that all the animals were in grave peril and it was my fault. In most scenarios, I realized the error of my ways when it was too late to make it right. One night, a parrot's feet fell off from neglect. He couldn't stay on his perch because, below the feathers, he had only stumps. I tried to glue his feet back on to no avail.

Before that, it was fish. I forgot to condition their water, and they immediately started floating to the top of the bowl. Then the tanks tipped over and there were fish everywhere, gasping for air. I couldn't get them back in the water fast enough. In one dream, hamsters were getting crushed beneath their exercise wheels. As soon as I got everything upright, the lid fell in and pinned the hamsters beneath its weight. In yet another nightmare, I was in a haunted house one night and stabbed a corgi with a fork before realizing he was my friend and would lead me to safety. I couldn't get the fork out, and he bled to death.

These dreams made my job a twenty-four-hour ordeal. During the day, I tried to avert normal disasters, but it seemed I couldn't ever be prepared or creative enough to anticipate the amazing array of mishaps that could occur. So at night, my subconscious went wild with the possibilities. I battled the inconceivable. Just as other people dreamed of missing a final exam or showing up to work without pants or forgetting they had a baby, I dreamed about killing hamsters and neglecting dogs; killing dogs and neglecting hamsters. As yet, the dreams didn't include domestic disasters within the households I was staying. I was sure it was just a matter of time.

On the second night, I planned to leave Charlie's crate and my bedroom door open. That way, he could get up in the night and let me know when he needed to go out.

Or so I assumed.

My sleep was fraught with the sound of toenails on hardwood and agitated yips from the hallway. I got up some time in the night, stiff-limbed and off-balance, to take Charlie to the garden for a potty break. I stood in the courtyard in my trout-print boxers and a tank top, teeth chattering while Charlie posed in the moonlight and stared at me. He didn't pee. He just stood there.

I brought Charlie's freshly laundered bed into the guest bedroom. If he was in the room with me, he'd have nothing to distract him from the seemingly simple task of sleeping. If he had to go outside, he could let me know with a polite nudge, or maybe a lick.

Instead, I woke up to Charlie's big black nose in my face, snuffling little flecks of dog snot into my mouth and—when I reluctantly opened them—my eyes. I only resorted to this scenario as a last-ditch attempt at all because Charlie, for all his many charms, had another habit that was distracting, to say the least.

Given his unusually protein-rich and dairy-laden diet, it shouldn't have been any surprise that Charlie was a stunningly gassy animal. His MO in my short time with him had been to fart and then leave the room. The smell was so potent that I had to leave the room also. No big mystery, then, that when he was in the bedroom with me, he gassed us both.

Sleeplessness seemed inevitable.

It was back into the crate with Charlie. On the third night, we went out again before bed, and instead of trying to engage in any kind of walk, I just waited. And waited and waited. At last, Charlie found something of interest to sniff on the monkey grass by the mailbox and finally delivered.

In the morning, he was curled on the hard black plastic molding of his crate, the cushion shoved to the front against the door. It was soaked.

To the Laundromat I went, his bed in a trash bag I found under the sink. It was a scented garbage bag, of course, and so sturdy you'd think it was made of stretchy fabric instead of plain old plastic.

That night, I gave Charlie free rein of the house, leaving my door firmly shut and the kitchen door to the patio cracked so he could get in and out as he pleased. I'd brought my own pillow from home to cover my ears, blocking out the *tap tap tap* of his nails as he trotted through the house.

And, glory be, it worked! I slept, deeply and uninterrupted, though I had another nightmare. I was supposed to keep two dogs separated—one on the porch, one in the house—or else they would kill each other. Then the maids came to clean the house and opened the screen door. By the time I figured out what had happened, the dogs were beyond repair. I had to send them downriver, like those Viking funerals where they float the corpse out onto the water in a boat or atop a twiggy raft.

In the morning, I couldn't find any evidence that Charlie went out in the night. But neither could I find any evidence that he went to the bathroom anywhere in the house. So long as he wasn't wetting the bed, though, I was growing less and less concerned about the regularity of his BMs.

On one of our limited morning jaunts around Katherine's front walk, I met a neighbor picking up her morning newspaper. She was embarrassed because I caught her in her pajamas. She chatted with me in a cryptic but very friendly manner: "How is your work going?" "I love your new shutters!"

Finally, in response to my puzzled silence, she said, "You're Katherine, right?"

"Oh, I am just the dog sitter. Katherine is in Italy."

"Oh, god! I'm sorry. You're both so tall, and with short hair. I thought . . ."

I was flattered that she'd mistake me for the elegant Katherine. But perhaps this neighbor in her cowboy-print pajamas, standing in front of a multimillion-dollar mansion with a newspaper in her immaculately manicured hand, did have cause to be embarrassed after all. How do neighbors living across the street not even recognize each other?

At the other end of the leash was Katherine's best friend, and wherever she was—probably staring at a relic from antiquity—she probably wished that this fickle, gassy beast were there at her side. For the first time, I saw her life with Charlie as isolated and potentially very lonely. Like Katherine, I had a single candidate for companionship, and he had four legs and a collar—a scenario that was all too familiar to me of late.

Once I reconciled that Charlie was my best and only antidote to loneliness, I tried to buddy up to him. He wasn't having any of it. Given my inability to pamper him in the manner that Katherine seemed expert at, Charlie largely ignored me and got even more cantankerous as our time together wore on. He ate less but farted more. I imagined him cutting his eyes meanly at me as he left the room in the wake of his outrageous flatulence. He spent most of his time prostrate in the corner of the living room, dangling his head at an unnatural angle out of his bed, his tongue lolling.

Feeling rejected and bored, I attacked the four-month-old stash of Halloween candy in the freezer, tucked in the back behind the rows of dog treats. I carefully rearranged the remaining candy to disguise the large dent I put in it. It was the only thing in the house to eat, except of course for Charlie's cottage cheese and hard-boiled eggs. I'd already eaten as much of the peanut butter in the cabinet as I dared without it being obvious. Of course, I could grocery shop for myself and keep food in her fridge, even cooking my meals in her fully functional kitchen. But, aside from avoiding further ways I could destroy her perfect house, I was also trying to save money from my narrow profit margin to eventually get an actual apartment.

As it was, crashing with my mom's best friend and her family on those nights that I didn't have overnights was undeniably convenient and economical. They were very generous to let me use their attic bedroom, and I was more than happy to help them out with domestic errands and the like in repayment. But six months of nomadic overnight nannying interspersed with what amounted to semi-permanent couch surfing had me longing for a room—or two or three—of my own.

When I went to work in the morning to do my usual rounds, Charlie refused to go outside to the garden. Instead, he ran around the house, dodging and barking shrilly when I came close. No amount of wheedling, bribery, petting, or kissing seemed to win Charlie over to my favor, and I suppose I deserved it.

It's not like he was the first animal I hadn't seen eye to eye with. Most recently, there'd been Cha Cha, a Chihuahua living in a duplex occupied by her owners—a couple, each living on their own side. Two people in a relationship living next door to one another and sharing the dog between them. That should have been the first flag.

Throughout the entire meeting, convened on Her side of the duplex but attended by Him as well, the tiny dog barked viciously and lunged at me, baring her miniscule but very sharp teeth.

"Oh, she is just protecting us," they'd both blustered. Cha Cha was little bigger than my foot, and I could've easily punted her through the living room window. The intensity of her apparent hatred for me was unnerving. I was trying to assure the clients of my competence but couldn't help jumping visibly at every bark. Any sudden movement on my part elicited a pointed snapping of her jaws, alarmingly close to my exposed wrists. I took very few notes during that meeting and tried not to move any more than I had to.

I managed to remember, however, that I was to come in through the back door of Her apartment, where I'd find a plate of cut-up hot dog. I would microwave that, tantalize Cha Cha with the treat, and hook her up to her lead for a walk while she was distracted by the hot dog pellet.

Upon my first—and only—visit with this pint-sized killer, I did manage to enter through the back door and microwave the hot dog. I attempted the distract-and-clip maneuver earnestly, offering the hot dog with one hand while aiming to hook the leash to the tiny D-ring at the back of Cha Cha's neck with the other. But that was met with more barking, bared teeth, and growling. Failing that, I tried tossing the hot dog and then sneaking up behind her with the leash. It all ended when she bit me and then peed all over the floor.

As Katherine's return approached, I got paranoid that she would psychically know about the candy, the washing machine, and everything else that had gone wrong during my stay. Even at twenty-two, I still lived in fear of "getting in trouble" or being accused of not doing something to perfection. For as long as I could remember, I placed unrealistic expectations upon myself—to be agreeable, amusing, unobjectionable in all ways. In school, an A– was not acceptable to me. The scholarship I received in college wasn't the best available, and thus I deemed it a failure. This pressure and these judgments came from me alone, and not my parents. They rejoiced over my one B grade, reading it as a sign that I was normal. They celebrated my fallibility, hoping this counterintuitive form of encouraging mediocrity would help me relax.

I recognized in Katherine a similar demand for excellence. Though where I strove, she seemed to succeed. Her standards, in fact, far exceeded mine, for I felt keenly while I was in her home with her dog that I fell obviously and irredeemably short of good enough in everything from having to use a towel to protect her bed linens from my zit cream, to my inability to bond meaningfully with Charlie as she had.

Rationally, I knew that if Katherine fired me, I wouldn't have my business license revoked or go to jail, and I wouldn't be heartbroken to never see Charlie again. But it really chafed me that I didn't nail this assignment. Even if Katherine couldn't tell—even if she never knew what went on in her house those days that I was there—I knew. I'd done everything expected of me where Charlie was concerned, yet I still had the distinct feeling of failure.

I spent a lot of time worrying about the tiny hairless patch left by the tree sap I'd yanked out, too. I saw a *Beavis and Butt-head* episode once where they gave themselves beards of head-hair, trimmed and then glued onto their chins. I thought that maybe I could trim some of his haunch fur, where it grew thicker, and glue it to the bald spot.

I wondered if three to four days' time was enough for Rogaine to take effect on a dog. But then I was afraid it would have some reverse effect on dog fur and make the bald spot bigger. I brushed the area compulsively, trying to arrange the surrounding hairs to mask the little spot, like a canine comb-over. Late into the evening, I stood across the room, and then closer in two-foot increments, trying to determine how close one had to be to notice the missing fur in Charlie's otherwise-perfect coat. He definitely looked thinner and was sleeping more. Or was he sleeping less and eating more?

Either way, it was clear to me that my isolation from people and near-exclusive interaction with dogs was taking its toll. I'd realized how singular the focus of my life had become at Christmas, during a visit from my parents. They'd occupied the pullout in Annie's den for the better part of a week, visiting with the family while I traveled hither and thither, my schedule crammed to overflowing due to the high volume of out-of-town clients over the holidays. I was beyond touched that they'd foregone their own deeply entrenched

Christmas traditions to be with me. I made a portion of sauerbraten, marinating it for the minimum three days beforehand, so my dad could have his usual yuletide dinner. On Christmas Eve, before I departed for my final pet-sitting visit of the night, we'd upheld the long-standing ritual of reading "The Night Before Christmas."

Christmas morning, I stole a few short hours to open presents with them, and every single gift was dog related. I received a pair of "Up on the Woof Top" socks depicting a dog-Santa dropping presents down the chimney; a CD carrying case in the shape of a dog's head, his mouth the zipper; a matching glow-in-the dark leash for dog and handkerchief for me (safety was a big thing for my parents that year, and always); pajamas from my sister and brother-in-law, sent with my parents, that read "Sleeps with Dogs" across the chest; a dog-themed address book, the dogs themselves illustrated as gumshoes, society molls, thespians, and roughnecks; a plastic purple poop-bag dispenser in the shape of a dog bone; a coffee-table encyclopedia of dog breeds; yet another subscription to *Bark Magazine* (my third); and a dog-paw-print "From the desk of" notepad, my name misspelled with an *a* instead of an *e*. My mom was profusely apologetic about that, claiming the manufacturer misread her personalization.

It was quite a haul. For a brief and surreal moment, I wondered that I hadn't been given any dog treats or a chew toy, as though I myself had pulled a canine metamorphosis and turned into a dog over the past five months. Other interests and dimensions, and certainly some human companionship, seemed immediately in order, lest I should actually start growing fur and barking.

Between the dogs by day, the occasional pet sleepover, and otherwise living with Annie's family and helping out with her boys, I wasn't meeting many people. And by people, I mean like-minded—in age or profession or hobbies or otherwise—individuals that could potentially be my friend. I really wasn't being picky. A human with whom I could recreationally share a meal, or a laugh. Or, if I was lucky, both.

I had a guest pass to Annie's tennis club and tried to get to as many yoga classes as I could. I found that the stretching helped ease the back pains I got from pounding miles of pavement during my days. The gentle yoga I liked was almost exclusively attended by women in their sixties, all of whom I grew very fond of, but none of whom seemed to be in the market for a twenty-two-year-old SWF BFF new to the area and craving companionship.

If I was expecting to find likely candidates at the monthly association meeting of area pet-care professionals, I came up empty handed there as well. I was at least ten years younger than the youngest of the other business owners, but proximity of age was less material in my friend quest than overcoming the distant yet professional cordiality that was the norm in the group. We were, ultimately, competitors in a fairly tight market.

I could tell early that there were alliances between business providers—walkers that helped each other out, or greatly respected each other's integrity. But this wasn't an environment ripe for grabbing a drink afterward, or making plans for the following weekend. Nor was it the place to gripe about particularly quirky clients, or tell stories on ourselves, admit recent screw-ups, or air our dirty laundry. Everyone was looking for a small edge on the others, and mistakes were remembered, catalogued quietly, and judgmentally held close. An ill-mannered dog that didn't improve over time could tarnish your reputation; if you took on a client that handled their pet irresponsibly, somehow that irresponsibility transferred to you. Everyone knew who'd been sued, or who would surely get sued any minute now; who ran chronically late; which walkers took more than the maximum number of dogs out on the trails; who charged too much or too little.

So I minded my p's and q's, tried not to ask too many questions, always complimented that week's snack-bringer on their baking or their choice of chips, and assiduously logged the meeting minutes, which I emailed out to the LISTSERV no more than five days following the meeting. I didn't want to gain a reputation as a procrastinator. Or an incomplete-minutes-keeper. I already felt that my age and relative inexperience were dings against my otherwise acceptable reputation within the group. And, ultimately, I accepted that I was going to need to look further afield for that much-needed human companionship.

I was out with Charlie on our most successful walk yet when my phone rang. Normally I didn't pick up when I was out with a dog—trying to have a conversation simultaneous to walking distracts me from noticing cars, other dogs, and countless other factors that could become problematic for me or the dog. Having one hand occupied with a phone and the other with the leash makes picking up a hot fresh poop nearly impossible. Charlie had reached the outer limit of his comfort zone at the end of the block and was poking around in a patch of ivy, hopefully getting ready to do his business. We could hang out there for a minute, so I answered.

It was a friend from my undergrad years who was finishing up his master's in English literature. He'd been sending me his short stories, hugely entertaining fictions about small-town giants, cadavers, hoarders, and misfits. We had taken writing classes together at school, and it seemed like he had more than enough talent and motivation to formally pursue publication.

The sprinklers were on in the neighborhood yards, and the sky was twilight purple. I was trying not to spook Charlie with my telephone conversation and jinx my unbelievably good luck at getting him past the mailbox and down the street, so I was replying to Ian quietly, and in short one- or two-word sentences. I still wasn't quite sure why he'd called. Usually he just emailed when he had a new story to show me.

While Charlie finally took a squat in the ivy, Ian was filling me in on his recent outbreak of psoriasis from the cold weather and the stresses of school. He said he'd stopped shaving and cutting his hair. He wanted to know what I thought of the last story he sent. I hadn't read it yet.

"So anyway, I gotta get out of the Northeast. My plan is to move out there at the end of the semester. I think we should live together."

In a way, we'd lived together before. During our senior year of college, we lived in separate apartments in the same building, a converted train depot that had seen better days. He'd explode into my loft without prior warning, pushing through the heavy ten-foot door, bellowing, "Hello the house!" Sometimes I'd be just out of the shower and standing in the hallway in only a towel. Or once, he came into my bedroom at three in the morning because I wasn't answering my phone. But I tolerated it. Sure, he had boundary issues, but he meant well.

We ended the conversation with him declaring he'd be out by the end of May, just as soon as he graduated and packed. I didn't believe he was coming.

In an effort to right all of my supposed wrongs, I brushed Charlie's teeth. I crept behind the bushes in the garden, scraping up the hard, heavy dog piles. I sat on the floor in the corner, rubbing Charlie's belly for thirty minutes at a time until my hand felt like corduroy. I ran around the house playing keep-away with him and his teddy. I hoped that my exuberance would dispel any angst or bad juju that was hanging over the house. I fluffed the candy in the freezer, made perfect hospital corners on the guest bed, refolded and returned the beach towel to its place in the back of the linen closet, and gave Charlie his beef-jerky-and-cottage-cheese treat.

I put the key on the dining room table before I left for the last time, along with the invoice and a note in which I baldly lied about how much fun Charlie and I had together.

In the aftermath of those overnights with Charlie, I'd started looking at studios in Berkeley—tiny affairs with kitchens that doubled as the living room and dining room, with walls so close I could almost touch both when standing in the center of the room. One had a slanted ceiling low enough that I had to stoop throughout the entirety of the tour. Still, I didn't think I could afford anything without having a roommate to share the cost.

I was seriously contemplating a place I'd dubbed "the phone booth"—a two-story converted storage shed with a two-burner stove top, a dorm fridge under the counter, a closet-sized bathroom, and a place beneath the stairs for a smallish chair—when Ian called again.

I'd been so certain he was bluffing, or that he'd flake on moving out to California. Yet there he was on the line, saying, "I can't think of anyone I'd rather live with, or any place I'd rather be. You can teach me how to cook! We'll have a book club and exchange our writing for critique. We'll be writers, and write together in our writers' den. We can find Michael Chabon and invite him and his wife to dinner!"

"So does May work?" he asked again. "Should I book the ticket?"

I was an easy sell, starved as I was for human contact and companionship in light of my professional sequestration with the animals I cared for. I was, after all, on the verge of renting a shoebox I couldn't afford, in which I couldn't even entertain without asking my guest to sit on the toilet lid or my bed. Ian's enthusiasm was infectious, and I even half-believed we might do one or some of the wholesome, self-actualizing activities he'd described. I said, "Sure."

The next time Katherine asked me back to watch Charlie for a week, I was staying with a Burnese mountain dog and his three-legged cat companion. The dog was as affectionate as Charlie wasn't, frequently clambering up into my lap as though he were twenty pounds and not eighty. He was a rapscallion, his puppy naughtiness lingering long past puppyhood. He loved to take his leash, or the edge of my shirt, in his mouth during walks and galumph off, dragging me in his wake. Worse, he chewed everything he could get his mouth around—napkins, the remote, car keys, socks, and, unfortunately for me, my eyeglasses. I'd returned to his house one night to find some kind of plastic twig dangling from his maw. I pried his jaws open to extract the mystery snack and pulled out the mangled remains of my frames. The lenses had long since popped out, and the rest looked like he'd been grinding on them all day long.

I couldn't drive, especially not at night, without my glasses, and they had to be replaced immediately. The owners wouldn't contribute to the cost, either, because I had known in advance that he ate everything, and I'd failed to put my glasses far enough out of reach. Raw deal, and really bad timing, as all of my extra money had been going toward the imminent move.

Whatever I perceived Charlie's and my differences to be, he wouldn't have eaten my glasses. When Katherine called, I had a sudden and unexpected rush of gratitude for him and his comparatively dignified, non-meddlesome comportment. I would have taken him on for another week, maybe even making friends with him this time, had I been available. But the dates in question conflicted with the week I was taking to drive from Georgia to California with my mom—a necessary trip, but a major ding to the old bank balance.

Moving out of Annie's house meant leaving their loaner car behind, a complicating factor of my transition from their attic to my own abode. Annie had of course offered to give it to me for next to nothing, but my parents had leapt at the opportunity to restore my old car to my care. They had no use for the extra set of wheels, and the car itself was a far better fit for this line of work than the Volvo sedan had been. While old, and fitted with manual everything, the car I'd be retrieving was a hatchback with a fold-down backseat, making it ideal for ferrying dogs about as needed.

Overjoyed at the prospect of a road trip with her youngest daughter and East Coast defector, my mom had planned the cross-country journey. I'd fly home, and we'd make our way back by car. She was also using the mother-daughter expedition as an excuse to return (or in her mind, get rid of) the rest of the furniture and boxes I'd left in their basement when I'd flown to California the first time.

Soon after I'd returned to California, mother and car in tow, and was setting up house in the apartment I'd rented in anticipation of Ian's arrival, I heard from the colleague who'd covered Charlie. Katherine decided to give him to her mother down in Orange County. He was apparently "too much trouble."

I was more than a little surprised to hear this, as I'd understood them to be blissful in each other's attractive and exacting company. Maybe it was his farts. Maybe she just couldn't deal with the smell. But she was the one feeding him a kilo of dairy a day. And if it wasn't that, what other aspects of his being was she dissatisfied with? In what ways was he inadequate? The toothbrushing, the teddy, the homemade treats; all of these were her doing, imposed on an animal that at the end of the day probably just wanted a good hard butt scratch and a pat on his bony head.

Hopefully that's exactly what Katherine's mother was giving Charlie down in SoCal. And maybe Katherine felt her life was improved, and her house more to her liking, without his crate next to her bed, his basket of accessories in the laundry room closet, his treats cluttering the freezer and his hardboiled eggs hogging up the egg crates, his dog house in the courtyard, his tartan cushion in the corner of the living room. But it had to be even lonelier there now, with just her gingham bedding, monogrammed towels, and mysteriously malfunctioning washer.

Of course, as I puzzled over my gross misreading of what I'd perceived to be a symbiotic relationship between Katherine and Charlie, I couldn't have known that I was about to learn for myself that living with someone isn't always a cure for loneliness. It can even make it worse.

To: "Mom"

Subject: Measuring up

Mamasita!

Thanks for the newspaper clipping. It is comforting
to know that animal-lovers across the country are as
crazy as they are here. I can honestly say that none of
my clients has a height chart for their dog. Although
some of the dogs do have insurance policies, private
rooms, and a wardrobe nicer than mine.

I know you always say that I didn't need to move out
here to prove anything, but—despite my immense
and ever-present homesickness—I think it has been
good for me. With everything I am learning out here
on my own, pretty soon I may even be able to say no.
Now wouldn't that be a trick? And, no, to answer your
question, I have not started reading *Codependent No
More.* But thank you for the suggestion.

Your definitively codependent daughter,
Lindsey

CHAPTER SIX

Alpha Females

....

Dusk was already falling, the sprinklers clicking on in the front yard with a hush, when I arrived at the stately mansion nestled among the quiet, wooded hills of north Berkeley. The Benjamins lived below street level, the garage set atop the multi-story house like a hat. Their dog, Felicity, always waited for me on the love seat by the front window, the creak of the gate at the top of the stairs announcing my arrival.

From her perch, she watched me descend through overgrown rosemary that flanked the stone steps. She remained there until the very last moment, body tense, moaning gutturally with anticipation. As the key turned in the lock, she was off, bounding around the corner to greet me with an exuberance unmatched by any of the other dogs I cared for.

Even after months of walking Felicity, my heart leapt a little when this eighty-pound German shepherd came at me. Her enthusiastic welcome—massive mouth open in a dog grin and fist-sized paws pounding the hardwood—would be truly frightening if I didn't know her so well. Rationally, I knew she would never do me intentional harm. She loved me. She was generally quite fond of all humans. But my knee-jerk reaction was to turn my back and protect my neck.

When I spent the night at the Benjamins', I traveled light. Unlike some of the other clients, they always had food for me in the refrigerator—leftover pork tenderloin, or a slice of pie baked in the oven of their beautiful Viking range. They kept the hot tub turned on and plenty of clean towels left out for my use. The DVD collection was vast (if a little heavy on the action flicks), and the guest bed was comfortable enough that I never needed to pack my own blanket or pillow. Nights at the Benjamins' were just fine.

I'd intended to arrive with enough daylight to walk Felicity, but my clients that Friday had run uniformly over schedule. Even with the later sunsets, heralding the coming of summer, it would be fully dark by the time I got Felicity all trussed up for her walk. In a neighborhood as nice as this one, I wasn't concerned about the criminal element. I doubted anyone would give me a hard time under any circumstance, tethered as I was to an animal as imposing as Felicity. Rather, when walking her, I needed the full visibility of our surroundings that daylight afforded for the safety of others.

Deciding her exercise would have to wait until morning, I dropped my bags and received the compulsory licking from head to toe.

"Let's see what they left in the fridge, *hmm*?"

Felicity preceded me down the hall and into the kitchen. The soaring ceilings were dark, the only light coming from the houses visible through the bank of windows at the back of the house, which overlooked the hills. I hit the switch and flooded the adobe-tiled room. The sub-zero fridge gleamed in the far corner. I fed Felicity first, filling her bowl with a heaping portion of medicated kibble. Felicity, like many German shepherds, had a sensitive stomach and either needed chicken breast with rice or the outrageously expensive dog food equivalent.

In the fridge, I found a gallon-sized Ziploc with a note:

Here is some fried chicken I made this week. There is tiramisu in the Tupperware. Help yourself. And thanks!

I'd lived my whole life in Atlanta before moving to California at twenty-one, and I still appreciated few things more than homemade fried chicken. Not bothering to heat it, I sank my teeth into a thigh, savoring the exquisite comfort of peppery skin. I thought of my dad and the proud way he showed me how his mom had always fried her chicken. "Pepper is the key," he'd say, dumping half a container into the paper bag that held the flour mixture.

Felicity finished eating before me and sat at my feet, expectantly waiting for a crumb to fall on the floor. When I was finished, I let her lick my fingers clean. I threw my paper towel in their industrial-strength compactor and went to put on my bathing suit.

It was during these visits with Felicity that I best understood the function and value of the overnight stays. Dog and cat hotels hadn't hit the big time yet, and boarding your dog or cat at the vet didn't closely resemble the customized, luxury pet vacation that is now the standard. Our poor dog would get so stressed when we put her in the kennel, she'd come home pounds thinner, her eyes gluey from lack of sleep, with a kennel cough that started to feel synonymous with Grant family vacations. If we knew of any live-in pet sitters back then, we'd likely have become faithful customers. Even though we had an outside-only dog, she'd have been able to sleep in her familiar doghouse, and gotten walked regularly, and not had to suffer the unfamiliar and likely uncomfortable enclosure, the other dogs' ceaseless barking, and the unnatural fluorescent lighting that didn't closely emulate the sunshine and fresh air she was so accustomed to.

While the standard of living at vets and kennels had surely improved since Biscuit's day, and pet hotels were starting to become more common and affordable, I still hated to think of Felicity being cooped up in some impersonal pen, getting perfunctory walks. She'd languish there! I was so glad there was someone like me who could come and share her bed and tell her how pretty she was. She deserved it.

Soaking in the Benjamins' aboveground hot tub was a strange mixture of anxiety and relaxation. The hot water and jets on my body felt miraculous, working away the soreness from days spent walking miles on unyielding concrete with dog after dog. The darkened yard where the sauna sat shielded me from any neighbor's watchful gaze, making my privacy complete. But no matter how much I tried to soothe Felicity or convince her otherwise, she seemed certain that I was in dire danger of drowning. She ran in circles around the hot tub, barking and whining, occasionally going so far as to walk up two or three of the molded plastic steps in order to save me from that mean, treacherous hot tub. I dangled my arm as far out of the tub as I could reach, stroking Felicity's head in an effort to calm her while the rest of my body floated weightlessly in the churning water.

The guest bedroom once belonged to the Benjamins' son, who had long ago left for boarding school and rarely came home for visits. From what I gathered, they went to visit him instead. Sitting on the hunter green bedspread, I gathered Felicity's massive head in my arms, kissing the dog between her gold-flecked olive eyes.

The first time I had ever stayed overnight at the Benjamins', so many months before, I mistook the master bedroom for the guest room. I'd marveled at the king-sized bed and the fine furnishing, assuming they took their guests' comfort very seriously. And I'd slept soundly, with Felicity curled at my side on the ample mattress. The next morning, I explored the house further and found the actual guest bedroom tucked under the stairs that led up to the garage. Embarrassed at my trespass, I stripped the master bed of its sheets and carefully removed any strands of blond hair from the decorative pillows and silk-embroidered duvet. When the sheets were clean, I tried to reassemble the bed exactly as it had been before I slept in it. I hadn't paid very close attention, having assumed I was meant to sleep there. I'd felt like a Goldilocks, moving my things from the gorgeous "too big" room into the "just right" guest room under the stairs.

Sleepy after my soak in the hot tub, I climbed between the nubbly flannel sheets of that "just right" bed, familiar to me now after so many sleepovers with Felicity, and fell almost immediately to sleep.

Nights at my own house hadn't been so restful of late. I am not sure what I'd envisioned living with Ian might be like. I think I'd fixated on the bliss of having my own bedroom, an ashram of sorts that was filled with the comfort and familiarity of my own things and not a stranger's. The roommate component was a means to that end. I hadn't taken into account the reality of the arrangement: that we'd actually be sharing the kitchen, living room, and bathroom, and that we'd be in constant negotiation over the use of those spaces and the items within. Or that he didn't have a job lined up when he moved, and still hadn't found one.

On the nights when I slept at the apartment, I'd lie awake, the TV turned up so loud that the mirror on the shared wall between my bedroom and the living room vibrated. He turned the volume up so he could hear his shows from the front stoop where he chain-smoked, a habit from college that had intensified for him in grad school, while I'd dropped my social smoking out of necessity when I started dog walking. I had no social life into which that casual compulsion might fit, anyway.

Of course, when I emerged from my room to ask him if he could turn the TV down a bit, or not smoke with the front door open, he'd silently comply. The next night he'd crank the volume again, and the smell of smoke would seep beneath my door, and I'd steel myself to ask him once more, hating myself for seeming like such a nag. Hating him for being so obstinate. Quickly tiring of this passive-aggressive do-si-do we danced every evening.

I never saw him before I headed out for work in the mornings, but evidence of his presence abounded. A row of beer cans decorated the kitchen counter like trophies, pubic-hair tumbleweeds scudded across the bathroom floor, his psoriasis creams crowded the lip of the sink, and the fug of cigarette smoke lingered, especially in the bathroom, where the vent fan had inhaled his prodigious nocturnal exhalations. Crack-backed, dog-eared copies of his well-loved books lay spread across the coffee table and couch: *Him with His Foot in his Mouth, The Berlin Stories, Lucky Jim*, bound to be replaced in coming days by another selection of cloth-bound, hardback first editions that lined the sagging bookshelf he'd claimed from the curbside in our neighborhood.

I'd been on my own, isolated from normal interpersonal inter-action, for long enough that these idiosyncrasies of his felt far more intrusive and intentionally aggravating than they should have. Beyond my constant requests to lower the volume and smoke well away from the door, I didn't know how to convince him to change his behavior.

Thus, the pristine silence at the Benjamins' house—and Felicity's non-verbal, non-smoking companionship—felt like a welcome respite from my new reality.

Later that night, I woke in a claustrophobic panic, sweating and unable to breathe. I was trying to escape a dream of being nailed to a board covered in coarse hair. As my eyes adjusted to the light and I got my bearings, I realized that I was face-to-face with Felicity, who had me pinned in a lovingly aggressive straddle.

"Get off me," I groused, weakly shoving at her. She rolled clumsily off and curled at my back, one heavy paw thrown over my neck.

In the morning, she leaped onto the bed with enough force to push the mattress slightly off the box spring. I turned away from her, blearily contemplating a breakfast of strong coffee and tiramisu in my immediate future. The Benjamins' Cuisinart coffeemaker, when filled with the magic of Peet's freshly ground beans, produced an irresistible reward for getting out of bed. So long as I didn't mistake the salt for the sugar. They inexplicably kept their salt in a porcelain bowl by the range, and I'd been duped more than once into ruining my coffee with a heaping spoonful.

After feeding Felicity, I opened the kitchen closet, where the dog's various walking equipment was stored—"her jewels," Mrs. Benjamin called them. Because she'd missed her walk the previous evening, Felicity was wired. Her eyes were locked on me in anticipation, her butt wiggling with excess energy.

"Sit!" I said in my most commanding voice.

First, I hooked the pinch collar around Felicity's neck. Then came the black vinyl muzzle, which slid over her snout and hooked around her erect, velvety ears. Finally came the shock collar, positioned around her neck so that the small rectangular box that delivered the shock was located at her throat, the most vulnerable spot.

The muzzle and the shock collar were recent additions. With all the straps and chains, I thought she looked like an S&M dog; all she was missing was the ball gag. As much as I hated the concept (not to mention the use of) the shock collar, I understood why it had been added to the armor. I didn't necessarily agree, but that seemed immaterial.

Felicity, trussed up and cinched into submission, walked down the hall in a tight heel, her head adjacent to my left thigh. At the front door, we paused.

"Sit!" Felicity sat.

"Heel!" She heeled.

We proceeded up the stone steps and out the gate onto street level, waiting for the steady stream of SUVs and convertibles whizzing by to subside. Finally a blond in a Jetta stopped, allowing for us to cross and head up into the hills.

The irony of my decision to walk dogs one or two at a time, always on-leash—avoiding, I thought, about 90 percent of behavioral or liability issues—was that the only clients who signed up for the one-on-one walks had dogs that couldn't be around by other dogs. Dogs with behavioral and control issues. Dogs like Felicity. In some cases, the leashes even exacerbated the dog's aggression. By eschewing the more popular—and better-paying—dog-walking paradigm of managing many dogs at once, I'd unwittingly backed myself into specializing in difficult dogs.

I had to learn quickly how to make my voice loud and deep, and raise my shoulders and spread my arms to seem as imposing and in control as I could. At six feet tall, this was easier for me than some of my more diminutive colleagues. Dogs have to believe you are the alpha. My height helped, but my ingrained submissive nature and reflexive passivity was a hindrance. Becoming convincingly assertive, aggressive even, presented a steep learning curve. It wasn't that I was unfamiliar with alpha females, but I'd never ever been in a position—or had the inclination—to act like one.

I was born in the year of the dog and, according to the zodiac, was a Virgo. I was the quiet yin to my older sister's outspoken yang. Four and a half years my senior, she is a horse and a Taurus, respectively.

I was such a quiet infant that my parents feared I was developmentally delayed. True, I'd been born slightly blue, the cord wrapped around my neck, and my subsequent APGAR scores were low. It became clear as I grew older, though, that my still waters were not a result of any disability. I was just that low-key: an observer, a thinker, and distinctly not much of a talker. My sister did that part for me.

From the moment I came into the world, she assumed the role of second mother to me. She dressed me, washed my hair, and spanked me when I misbehaved—and when I didn't. She was preternaturally authoritarian and maternal, which meshed perfectly with my submissive nature. Or perhaps it was the catalyst for my acquiescence. Whichever way our complementary characteristics came to be, we were inseparable, the opposition of our temperaments creating one balanced child out of two.

I grew up oblivious that having two mother figures was at all out of the ordinary. Instead, I learned early which mother was good for what. My sister was a constant companion, but I was careful around her to not to do anything that might warrant punishment. She spanked me far more than my parents ever did. Contrasting with my sister's authoritarianism, my mom represented all things soft and warm and good. She rarely disciplined me, probably because she never really needed to. I was kept pretty well in line without her involvement.

My sister and I were fiercely bonded, my love for her always all-consuming, but mixed in with my adoration of her was a fearful respect for her dominance. She picked the movies, got first go at the front seat of the car, chose the restaurant, told the story, got the laughs. And in later years, she got the guys, too.

Even as an adult, I seemed to gravitate toward strong, bossy women, or they to me. For as long as I could remember, I had friends that called the shots—whether that meant I always played the boy in make-believe, or, in older relationships, that I was usually the designated driver or picked up the tab more frequently. It had never really occurred to me that there was a different way. That I might assert my needs or opinions a little louder; that compromise might be a nice solution for everyone involved.

Felicity's and my usual route crossed a second busy thoroughfare, the final main artery twisting into the more remote hills above. Sometimes we'd wait up to five minutes for a lull in the traffic, darting as quickly as possible to the far side of the street. It was my least favorite part of the walk, and not just because this was the spot in which "it" had happened.

A month before, we were heading back down the hill toward the house and paused as always at the crossroad. Waiting there, I had loosely straddled Felicity, my legs on either side of her wide back. At that time, she'd only worn the pinch collar and a gentle lead around her snout. I was stroking her head, passing the time until the unbroken procession of cars slowed long enough for us to cross, when I heard the patter of paws on pavement and a delighted-sounding yip.

I saw a flash of black under Felicity and realized that, in the space of a moment, she'd taken this little dog's neck in her jaws. A geyser of piss shot a foot in the air; the puppy was bloody before I could react. I pulled on the leash to the point that I was nearly sitting on the pavement, but Felicity could not be interrupted. She had the little black Lab by its neck, and she was shaking it. Hard. In my panic, I didn't think of the pepper spray in my fanny pack, carried for occasions like this one. I couldn't make myself big enough, or my voice loud enough, to make it stop. There was blood everywhere.

A man had come running toward us and reached into the fray, pulling Felicity bodily off his pet as if by magic, through some superhuman surge of adrenaline. The puppy, once released, went slinking back up the hill, trailing blood behind him. The man was panting. Felicity licked her lips and sat down.

"You'd better go check on your dog," I croaked.

It took the better part of a month to settle the fallout from the attack. Only one thing was sure: the puppy was off-leash and Felicity was not, so his owners and not the Benjamins were at fault. That is not to say that the owners did not want compensation for the vet bills—the cost of the stitches, the ointment, the antibiotic, and the Clomicalm for the pup's shredded nerves all added up to a hefty sum.

The Benjamins refused to contribute. This was awkward for me, as I had to pass the victim's house every time we went on a walk. And every time, the puppy barked from where he was tied up on the porch, Felicity tensing in what seemed like excitement or anticipation. The owner told me one day, as I tried to slink by the house unnoticed, that their little dog wasn't the same after the attack.

The subject of my own shredded nerves never came up; they didn't ask how I was feeling following the incident, and I felt ashamed to initiate a conversation about my anxieties. I thought that, by them not asking, it was assumed that I was fine—that I should be fine. That I could handle it. Oh, how I wanted to be that strong and imperturbable.

Guilty or not, Felicity's gentle lead was immediately discarded in favor of the muzzle and shock collar. Mrs. Benjamin demonstrated her new and distinctly not-gentle accessories the next time I arrived to walk her.

The muzzle fit snugly around her snout, preventing her from doing anything more than a pathetically restricted pant, the tip of her tongue protruding as much as the thick canvas sheath would allow. The shock collar came with a remote that I carried in my hand. It had a dial, one to ten, which dictated the intensity of the shock that I delivered via a large black button. Not something to press inadvertently. Not something I ever wanted to press intentionally, for that matter.

My slumber party with Felicity fell over a weekend, and I didn't have any other dogs to walk or pets to visit. Few weekends were as free as this one; usually I had a couple of drop-in visits for dogs and cats whose owners worked weekends or were out of town.

After our morning walk, blessedly devoid of any other dogs or deer or potential for disaster, I settled down on the back porch with Felicity at my feet and read a novel. I was just as happy to spend the Saturday in solitude; I presumed Ian had plans of his own that didn't require my presence at the apartment. No doubt he was busy populating the folds of our slipcovered armchair with spit-dampened sunflower seed husks.

In the afternoon, Felicity and I dozed in the den. I was curled into a tight ball on one end of the couch, Felicity in a heap at the other. Despite my best intentions to get her second walk in during daylight hours, we overslept, and the sun was already waning by the time she was strapped up and ready to go.

Walking up the stone steps, I recoiled at a giant garden spider weaving its web between the rosemary and a low-hanging branch of the Japanese maple. The mantra "There's nothing to fear but fear itself. And spiders" ran through my head as it always did in the event of a close encounter.

Still shuddering, I unlatched the gate. When Felicity tore past me, I fell and hit my head. From the ground I could see her shadowy form darting through the evening traffic, the headlights illuminating her race to the far side of the road. Only when I was back on my feet could I see what she was after.

The other dog's owner had dropped the leash, screaming shrilly as Felicity chased her golden retriever around the cars. Felicity seemed utterly unaware that the muzzle prevented her from truly getting at her victim; the retriever, on the other hand, got a defensive snap and a snarl in when it could.

"Control your fucking dog," the woman was screaming at me. I was weaving through the cars, which had stopped to observe the strange scene unfolding around them. As traffic started to back up, drivers farther from the scene began to honk impatiently.

I was pressing the shock collar remote repeatedly, the knob cranked to deliver the maximum voltage to the box around Felicity's neck. But she couldn't be distracted. She didn't even seem to register the shocks.

Around and around we went, me chasing Felicity, Felicity chasing the terrified retriever, the woman screaming like her dog was being gutted before her eyes. Finally, I got a foot on Felicity's trailing leash. My adrenaline gave me power enough to drag the charging shepherd to me, though I was shaking badly. It was clear that the woman's dog was unharmed, thanks to the stopped cars and Felicity's secure muzzle. But the retriever—like me—was trembling visibly, traumatized by the sudden and unrelenting pursuit of the bigger, overtly aggressive dog at my side. The owner appeared ready to murder me with her bare hands, as if it were me and not my canine companion that had given chase.

"I am so sorry," I stammered. "Is there anything I can do?"

"Control your fucking dog!" the woman spat over her shoulder, as she set off up the hill.

"I'm so, so sorry," I called after her, my voice trailing off at the obvious futility of my broken-record apology.

The cars had started moving again, now that the spectacle had ceased. I waited my turn to cross back over to the house, taking Felicity back through the gate, down the steps, and into the entryway where'd we'd been only moments before, when we were intact and still innocent of all charges.

Biscuit had been anything but aggressive—to humans, squirrels, birds, rats, or other dogs. My parents told the same tired joke over and over, that if someone tried to rob us, she'd probably lick the thief to death. The only things she snapped at were the bees that buzzed in her face as she dozed in the shade of our heavily wooded backyard.

When she was home alone, she had a long lead in the back that gave her the full run of the yard, but when we were home we didn't worry too much about her wandering off. We could always find her across the street with her boyfriend Zeus, a handsome golden retriever with a deep bronze coat and white chest. On one such afternoon, when she was unleashed and enjoying free range of our sprawling front yard, she trotted down the driveway to greet a dog walking by with his owner. The woman was pinch-faced and redheaded, and she was wearing an ugly coat. The dog lunged at Biscuit, baring its teeth. Biscuit's tail was still wagging when the owner took a rolled-up newspaper and smacked her over the head repeatedly. She took her own snarling dog on up the street while Biscuit slunk back to me with her tail between her legs.

When I removed Felicity's muzzle, I could see the retriever had gotten a few good bites in on her face. She was bleeding between the eyes and under her right ear. For my part, I had scraped an elbow and a knee, and a tender lump was rising on my temple, right at my hairline.

"All right, you unbelievably bitchy beast," I said wearily. Using a dampened paper towel and soap, I cleaned the puncture wound. Felicity was like a different dog, completely calm and licking my hands as I cleaned her up, as though I were simply caressing her face and the events of the previous ten minutes had never even happened. I dressed the wounds with Neosporin, wiping the residue on my jeans.

"What are your parents going to say about this?"

I dreaded that conversation, though I knew it was unavoidable. I'd been cleared of any wrongdoing or negligence in the case of Felicity versus the off-leash puppy, but it was still a strike against our collective record—Felicity's and mine, together as a dog-and-walker unit. I suspected that there were not three strikes in this game, and that the events that had just unfolded at street level would mean the end of these walks with my selectively ferocious companion.

And I was right.

After the second attack, I still spent the night with Felicity. We had naps on the couch, and I continued to share the guest bed with her. The Benjamins still left me food in the fridge, though I perceived a distance in their once-warm notes. This was likely projection on my part, as I anxiously looked for exclamation points, smiley faces, any indication that they still held me in high regard. But I never walked Felicity again, the consensus being that I was not strong enough or skilled enough to control her.

I was not alpha enough. All of these things were true, and I resented it bitterly.

Disgruntled as I was about my demotion, I went along with the new arrangement. It was less that I really wanted to be the one putting myself (and all dogs, large or small) in harm's way so that she could be well-exercised, and more that I hated to be deemed inadequate in any way. Especially in this way. In my most honest self-recriminations, I acknowledged that I was always the pushover, even in adulthood. Even when it came to dogs. Even when it came to roommates.

My parents—and everyone else I told about the situation— asked why in the world I didn't just walk away. The money, they said, couldn't possibly be worth the bullshit.

And they were right. The money wasn't really the point, ultimately, though I sorely needed the income from this client. The loss of her regular weekly walks had put a significant dent in my monthly balance, which forced me to refactor what at the apartment I could jettison in the aftermath of this pay cut. We didn't have cable, using old-fashioned rabbit ears on the ancient TV I'd found on Craigslist for $20. I was still relying on pirated Internet from the coffee shop by the apartment. I'd drive up and park curbside, taking care of all Internet-based business from the driver's seat of the car. There was little fat we could cut out of the current arrangement. I wasn't willing to compromise on power or hot water.

Money entirely aside, I kept up the overnights when the request arose because I couldn't abandon Felicity entirely. For better or worse, she was subject to owners who addressed her desire to munch other dogs to death by trussing her up and attempting to shock her into submission. I hoped, perhaps foolishly, that my love for her might in some small way offset the trauma of their approach. Sad as it was, she was the closest thing I had to a friend at that point, and I knew I'd miss her too much if I quit the client altogether.

The Benjamins enrolled her in a highly reputable—and appropriately expensive—training program for aggressive dogs. It was with an unbecoming sense of schadenfreude that I learned Felicity took off after another dog at one of the classes and pulled Mrs. Benjamin down, dragging her for a short distance and cracking two of her ribs in the process. Apparently she wasn't alpha female enough for Felicity either.

After that accident, she seemed to concede that Felicity's previous transgressions weren't my fault. She never said as much, but the rapport between us became markedly less chilly. I liked to think that she understood that no one—neither she nor I nor anyone else—could be expected to easily bring to heel an enraged eighty-pound dog that, despite her bedroom eyes and propensity for spooning, was still an animal. With all the accompanying natural instincts for hunting and establishing dominance. And resisting electrocution.

Perhaps, and this was pushing my luck, she finally understood that Felicity couldn't be controlled. Not by anyone of normal strength and a reasonable aversion to canine torture. If nothing else, we had something in common; we were both victims of Felicity, and her insatiable penchant for mayhem. And we both loved that dog immeasurably, no matter what harm she did.

In the end, it wasn't an alpha female who changed Felicity at all. It was Tom, my mentor and the one who had introduced me to Felicity and the Benjamins in the first place. He'd walked her long before I ever came on the scene, before the muzzle and the shock collar and our many misadventures together. He, too, believed in the power of love and a soft touch. But unlike me, he put his faith into action and blew us all away with his impossibly quick rehabilitation of our irascible bully.

After mere months under Tom's careful tutelage—without the use of a choke or pinch collar, and certainly not the shock collar, or any other form of negative reinforcement—Felicity was riding in his van with other dogs and romping about out on leash- and violence-free group walks. This, the same dog who had nearly killed a puppy, muzzle-jumped an innocent retriever, and mangled her owner's rib cage.

Thrilled as I was by this transformation, I'd be a big liar not to admit that I wished it had been me who brought our girl back from the dark side. I had the pure, unadulterated desire to make her happy. But actually achieving that required skills I simply didn't possess. At least not yet. The question was whether I could learn to be as assertive as I needed to be. And whether I could learn fast enough. My success in this line of work depended on it. While I wasn't yet an old dog, per se, it was a hell of trick for me to pick up after so many years of being carefully and comfortably submissive.

The alpha in me, it seemed, remained ever elusive, both in matters professional *and* personal.

To: "Sis"
Subject: Pizza-face thanks you

Sister mine,

I got my first shipment of Proactiv today. Thank you
for this! My face is just too awful to bear—I can't
even touch it, it's so bad. Truly a pizza face. Until
the medication works its magic, I am thinking about
wearing one of those *Phantom of the Opera* masks—
except I think my jaw would still be exposed. Maybe a
traditional paper sack would work best? I promise I'll
pay you back. Do you think I am allergic to dog saliva?
I'm probably allergic to having no money. Maybe the
pharmacist could prescribe me a roll of twenties . . .
I'm going to look into it.

Lindsey

CHAPTER SEVEN

Desperate Measures

....

My cell phone's musical ringtone jarred me from sleep. Cheerful as it might have sounded to the untrained ear, it had started inciting in me a physical response. A tensing, or a bracing, in recognition that there could be nothing but bad news on the line. This was likely due to the fact that for every kindly call I got from a family member or, even more rarely, an old friend, there were dozens coming in concerning this dog or that appointment that required quick thinking, flexibility, and light-on-my-feet troubleshooting. Hence my Pavlovian response to the ringing phone.

Considering that the sun was barely up and I'd been fast asleep, I was hardly primed for any feats of problem-solving prowess. Before answering the incoming call, I cleared my throat hard, giving my voice a few practice runs. I always sound phlegmy and hung-over in the morning, whether that's actually the case or not. Either way, I didn't need to give my clients the impression that I was answering the phone in bed, still half-asleep. Even though I was, on both counts. I didn't want the person on the other end of the line to free associate toward unbrushed teeth and rumpled pajamas.

"*Uhhhahhheeeeaaaaoooooo.*" I cleared my throat again. "Good morning. Good MORNing!" I flipped open my phone.

"Hello?"

"Good morning, hi! It's Sally, Tucker and Abbie's mom."

Their puppy profiles popped into my head automatically. *Tucker and Abbie, Border collies. She's purebred; he's a mix. She's high-strung; he's just hyper.*

"Sally, hi!"

"Sorry to call so early. I need to cancel Tucker and Abbie's walk today. By the time I remembered I should call you last night, it was way too late. I'll pay for the late cancellation."

"It's okay! No worries." I was way too nice. Cancellations were a pain and could wreak havoc on a carefully scheduled day of visits, but it was pretty cool when they were late enough to warrant payment. I felt bad enforcing the penalty, though, and rarely held my clients to it out of sheer codependency. Especially in this case, when Abbie and Tucker were my first visits of the day and didn't throw much else off schedule.

"Abbie's had a terrible accident. She jumped out of the living room window yesterday before I got home from work. I found the screen popped out, and Abbie was down on the patio just covered in blood. Even after the stitches and medication, she's in pretty bad shape, so I'm going to take her in to work with me for the rest of the week."

My prior experience with animal suicide was limited. When I was a baby, the family dog chewed all her fur off in a gruesome reaction to her extreme flea allergies. Dinah, hairless and grotesque to behold and still in itchy agony, no doubt, was euthanized soon after.

More recently, I returned home to the apartment in Berkeley and found my $2 goldfish, Chubs, lying on my bookcase, his dried-out gills still twitching. I put him back in the water, but he was a floater within minutes. I wouldn't classify either of these as attempts at suicide, although I guess my fish came much closer than our poor, tormented dog. I couldn't rightly say what was going through his fishy little mind when he jumped the bowl. Was it an accident? The result of overzealous exercise?

Luckily, Abbie's leap from the second-story window hadn't been fatal. Sally spared no detail in recounting Abbie's fractured rib and a large gash in her left rear haunch from hitting the downspout in the fall.

I was still trying to envision how this would have been possible. Sally's living room did have floor-to-ceiling windows, and I supposed, with the sash raised, it was of a height for a dog to leap out—screen or no screen. But what an odd sequence of events. I couldn't really wrap my brain around what Abbie's frame of mind must have been to do such a thing. It didn't seem possible that she'd inadvertently fall over the lip of the window and break through the screen. She and Tucker would have had to be playing one hell of a game of chase for her to accidentally catapult herself that hard and that far. I was starting to see the practicality of keeping a nanny cam after all; many a mystery might be solved with the help of a live feed, documenting the events of the dogs' every day.

"Abbie's vet upped her Clomicalm and added a prescription for Deramaxx for the pain, as well as a sedative if she goes into panic mode. I'm worried about Tucker as well—being alone in the house is going to be distressing for him as long as Abbie is coming to work with me. Let's touch base after his walks and stay in communication about his spirits, okay?"

Had I been more conscientious, or more awake, I would have

pulled the dogs' file from my cabinet and recorded Abbie's medication names and the prescribed amounts, along with her injuries and the implications for healing. For Tucker's "extra emotional support," I was making a mental note to give him some additional turkey treats and plenty of vigorous scratches under his collar. I'd worry about Abbie's pharmacopoeia when the time came. She'd be with Sally for long enough that her regimen of medications would likely change before she came under my care anyway. For now, I was happy to let Sally lead the suicide watch.

Of the two dogs, Tucker was a Border collie mix, where Abbie was of purebred provenance. I attributed Tucker's slightly reduced levels of crazy to this nuanced difference. He was still arguably hyper, but it didn't verge on unpredictably neurotic like Abbie's temperament could. I never imagined she was capable of self-harm, though, and her desperate leap left me not just surprised but also feeling very sorry for the poor girl. She was clearly suffering more than anyone knew.

With the first appointment of the day canceled and my alarm set to go off in over an hour, I flipped my drool-dampened pillow, curled into a ball, and drifted back to sleep.

I was noticing that many of my clients' dogs seemed to require, and certainly received, an inordinate amount of medical attention. These two were no exception. Just days before, Tucker had chipped a front tooth—how, I couldn't say—and Sally had it capped.

I hadn't been present for the chipping or the capping, but, in Sally's note to me following his dental procedure, she warned me that his anesthetic might not have entirely worn off and that I might keep him on-leash for the walk in case he was woozy. She went on about how impressed she was with the job they'd done, and how affordable it was thanks to Tucker's dental coverage. I was in awe that Tucker even had dental coverage. I could have used some of that, as I was entirely without health insurance of any kind since graduating from college the year before.

This was one of the very few lies that I had ever told my parents, and I knew it was a biggie. Not just because they wanted me to be well cared for, but because a medical emergency without any kind of insurance could be financially ruinous. I wasn't just gambling with my health; I was playing fast and loose with their savings. I was learning, too, that working with occasionally aggressive dogs didn't exactly reduce my chances of having an accident that would require medical attention.

Uneasy as I was with untruths when it came to my parents, who truly deserved nothing but the best and most honest communication from me, I avoided the topic of my health as assiduously as I could. Unfortunately for me, it was one of my mother's favorite means of checking up on me. She was forever asking how I was feeling and was I coming down with something? Was I sleeping enough? Eating enough? I tried not to clear my throat too often or sniff too loudly lest she jump to the conclusion that I had a cold coming on.

Whether or not this was the cause for my mother's near-constant concerns about my health, I don't know, but I'd been on antidepressants since the summer after my sophomore year of college. My nervous breakdown would have been, well, depressing, if the resulting prescription hadn't made such a miraculous difference in me.

I'd planned to spend the summer months as a counselor at the Quaker farm camp I'd attended as a child. When I arrived and started orientation prior to the campers' arrival, something about returning as an almost-adult to such a significant place from my easy and innocent youth, and assuming so much responsibility for these contemporary versions of little me, broke the barrier that had held my long-ignored depression at bay.

Instead of spending those long sunny days guiding kids in cow-milking and berry-picking, mud-wrestling and quiet contemplation, I returned home for an alternate summer filled with talk therapy and a new prescription for antianxiety meds. It was my good fortune that the camp director had long suffered from bipolar disorder and was exceedingly understanding in her acceptance of my sudden and inconvenient resignation.

Instead of feeling ashamed about my complete and rather public decompensation that notorious summer, I was more grateful that it was all so infinitely treatable. Everyone's depression manifests differently; for me, it was characterized by anxiety that was particularly pronounced in social situations. I blushed and sweated profusely, my glasses fogging up and my palms going slimy, making me even more self-conscious than I was to begin with. Loud noises and big crowds set my heart racing, and I was generally jumpy and had a hard time focusing. I always felt like I should be somewhere other than I was, and I generally didn't want to be seen by anyone.

But with twenty milligrams of an SSRI on my side, I was a completely different person. Though I was still an introvert in every way, I was able to hang out comfortably with humans. I even sought their company. I was only dismayed that the issue had gone untreated for so long; on antidepressants, high school might not have been quite so traumatizing.

Currently, without any insurance to speak of, dental or otherwise, sixty bucks for the off-plan Citalopram prescription was way steeper than I could afford. I needed that money for essentials like gas, and food, and poop bags, and dog treats. When I realized I could no longer spring for the suggested dosage, I'd been careful to wean myself off the meds. I couldn't imagine what going cold turkey might have been like; even tapering off was harder than I liked to admit. I could tell that my anxiety levels were increasing, the old symptoms quietly blooming.

Working with the dogs was a blessing and a curse when it came to managing my depression. I was relieved to remain in their non-judgmental company; with them, I never felt the urge to flee to the safety of the bathroom or car until the evening was over. Even if I did, they wouldn't know the difference or care either way. I'd long felt that animals were so much easier to read than my human counterparts, the energy of their temperament seeming to radiate off them with clarity. I could better understand their motivations and needs and was soothed by their easy, unquestioning acceptance of me.

Spending so much time exclusively with animals was also rendering what social skills I had rather rusty and was allowing me to retreat from the sometime-difficult but very necessary act of socializing and negotiating with my own kind.

If my mom's favorite inquiry was my health, my dad's was my finances. Another subject I was happier to stay well away from. When we'd chat, he'd often pepper the conversation with his tried-and-true nuggets of financial wisdom. "Save your money—you'll always need it." Or, "Try to put at least a quarter of what you make in savings." "Don't replace the air filter every time they suggest it," and "Change your own oil!"

I'd try as hard as I could to hear him and not get upset by how far behind the curve I felt when it came to being a responsible adult. I was working each month's earnings down to the penny, scraping and scrounging in every way I could. I was making meals out of the free samples at the Andronico's by my house, buying gas with quarters pilfered from Ian's change jar, and paying almost all of my bills late. When he'd ask if I was okay, if I had everything I needed, how was the car, and was I changing the oil regularly but not too regularly, I'd try to play along as best I could. As much as I wanted to succeed on my own sweat for my own gratification, I was equally motivated by a desire not to disappoint him. To prove I could do it all by myself.

When it was really bad—when I was days away from my next payment and dollars from overdrawing on my bank account, which I was doing more and more frequently—I'd occasionally crack and the truth of the situation would come out. And he always helped; he never even thought twice about it, so enthusiastic and proud was he of my entrepreneurial spirit and intrepid choice of career, however temporary it might be. He never minded giving me a hand from time to time. But I minded. I noted every dollar he lent me, even when it was clear that it wasn't a loan but a bailout. Alongside my ledger of miles traveled for business, gas spent, and supplies purchased, I logged dollars owed back to my parents.

Surely he'd done the math and suspected I wasn't making enough money to live on my own, with or without a roommate. It was increasingly obvious even to math-challenged me that I'd grossly miscalculated how much adulthood would cost, especially in northern California, and I wasn't sure how to bridge the gap between what I made and what I needed to be making in order for this to work out. The overnight visits were my biggest moneymaker, yet what I made in a night barely covered what I was paying per day in rent. Even I could see how little sense it made to lease a room I wasn't occupying, covering the rent by getting paid to sleep elsewhere. This constant and dire lack of funds did nothing to help my stress levels.

No matter how honest I got with my dad about money, and what I could or could not afford, or how well or badly I was maintaining the car, or how neglected or nonexistent my savings account might be, there was no way I'd ever confess that I was off my meds. I was sure that on his and my mom's parental seismograph, this would signal to them that I was in crisis and needed to pursue a different course. I had no doubt they'd board a plane as soon as possible to be by my side as we figured out what that course might be together.

I suppose I could argue that because I was working with dogs, and my exposure to intense social situations was thus lessened, so, too, should be my need for medication. If only the therapeutic effect of the animals' companionship was enough to regulate my anxiety levels. But chemical imbalances are rarely so easily addressed, mine included. In the absence of the antidepressants, at least I was getting plenty of exercise and vitamin D. Walking upward of eight miles a day had to help with the distribution of serotonin in my system. If only I were having some healthy, confidence-boosting sex, too. Or any sex at all.

But I could absolutely feel the difference, now that all traces of the fix-it-all pills had disappeared from my system. It didn't matter whether I was in a room full of people or sitting alone in my room—I was off-balance, easily distracted, often panicky, not myself. It was all I could do to mask any evidence of this when one parent or the other called to check in.

Abbie had been taking a baseline dose of the antianxiety meds—in her case, for separation anxiety—long before her bizarre leap from the living room window, and I'd wondered more than once how much her prescription cost and how the pharmacology of the canine version differed from the human. Could she also tell a difference on the meds? Did she feel more like Abbie than in her pre-pill days?

It would be some trick if I could save money and avoid the encroaching med-free malaise by taking an antidepressant intended for dogs. I already ate Abbie and Tucker's dog biscuits—the all-natural kind with normal human ingredients made bland enough for dogs. We used to eat them at the pet store, kind of as a joke, but kind of not. They were pretty tasteless, but definitely not bad. And they took the edge off when I was really hungry.

When my phone's alarm finally went off, trilling a different but equally nerve-shredding tune as the ring tone, I threw the covers back groggily. With my feet flat on the floor, I mentally reviewed my schedule for the day, now minus Abbie and Tucker. I'd work my way up 880 toward Richmond, where I'd visit Morgan the boxer mix, followed by Spence and Doodle in Richmond proper. All my late-afternoon visits—a drop-in visit for two cats, a neighborhood walk for a Lab, as well as one with a pit mix and greyhound—were in the exact opposite direction, back through the MacArthur maze toward Oakland and Alameda.

In the dead of summer, copious sunscreen and repeated applications were a must for me. Even with SPF 55 applied liberally throughout the day, my freckles stood out dramatically against my pale Scottish skin, giving the illusion of a mottled, uneven tan. In coloring, I was my father's daughter through and through, my reddish blond hair and supernaturally easy-to-burn skin the exact opposite of my mother and sister's more olive coloring. I wore a hat as well, learning the hard way that my scalp and ears were always the first things to roast.

Even though the canceled appointment freed up my morning significantly, I was sorry to miss the trip up to the top of the El Cerrito hill where Abbie and Tucker lived. We walked a path down the road from their house, an unpaved swath cut through knee-high grasses and gnarled trees on a steep hillside overlooking all of the neighborhoods that stretched toward the bay. The panoramic vista spanned the Bay Bridge over to San Francisco and included the always-majestic Golden Gate and the lush green hills of the

North Bay. In the foreground, BART cut through the urban sprawl like an electric model train set. From up so high, the cars seemed to crawl along 880 like little wind-up toys. On that walk at the top of the world, I could see the neighborhoods I'd visit next spread out below us: Richmond Annex and then the city of Richmond just to the north. Here, as anywhere, the neighborhoods got less desirable farther down the hill, and both of those later-in-the-day clients lived in the flats between train tracks and freeway.

Even as I enjoyed that spectacular view from the hillside, the walk itself posed plenty of challenges. In the dry summer, the insidious foxtails were all too plentiful. I lived in constant fear of these barbed seed pods that literally burrowed into dogs—through their nostrils, ears, paws—working their way ever farther into their system, causing excessive pain and requiring surgery to remove. It's asshole weeds like these that cause pet owners to get health insurance. One foxtail-removal surgery alone would make the policy worth it.

I'd already had to make a foxtail-related emergency vet visit with a dog I was pet-sitting recently. During our evening walk, much of it spent with the dog's nose to the sidewalk and plants along the verge, my charge started sneezing forcefully and continuously. This went on for the remainder of our walk, and I grew convinced that he'd snorted a barbed little bastard of a foxtail into his nostril. It turned out to be a false alarm, possibly an allergy to something else he inhaled. I apologized to the owners excessively for the unplanned vet visit and subsequent bill. But I—and they—figured it was cheaper to have a foxtail false alarm than to leave the real deal untreated and have to eventually pay for a more complicated extraction.

Beyond the threat of foxtails on this particular trail, there were plenty of deer to distract these already highly distractible dogs, as well as other dog walkers with hordes of off-leash charges. Tucker and Abbie were technically an off-leash walk—only one of two clients of this kind that I took on, given my liability terrors—but I always waited to assess the situation on the path before releasing them to walk free. Though we were up high and could see the surrounding cities without obstruction, the visibility on the twisty, hilly trail itself was tough. I never knew who or what awaited us around the bend, or if something might pique one of the dogs' interests and cause them to bolt. Abbie was especially inclined to take off for heart-stopping minutes at a time down the path or up an overgrown hill, for no apparent reason. One week with Tucker off-leash would be infinitely more manageable, and not just because the numbers were in my favor. For any off-leash walk, I much preferred the ratio of one me to one dog.

Abbie's dramatic and unexpected dive prompted me to acknowledge that most of the dogs I took care of were a little nuts. It was a wide spectrum, but I could place almost all of them somewhere along the range of dysfunction. At the least extreme end were dogs like dear Buster, on antidepressants for separation anxiety. Worse were Beano and Discus, two seriously yappy shelties in the Berkeley hills. When I'd arrive to collect them, they'd bark as viciously as if they were trained guard dogs ten times their actual size—and with none of their extreme and highly disarming fluff—and I were a gun-wielding, dog-hating maniac. Treats placated them for long enough to get them leashed up, but then, throughout their walk, they went into sudden and unexplained frenzies in which they both spun in tight circles and barked their tiny heads off. The mania eventually subsided and we could continue on. At this point on the spectrum, the misbehaviors were undeniably annoying and usually inexplicable, but relatively harmless.

Then there was Foxy, an Australian shepherd I walked in the

area. She was a lamb with humans—all toothy smiles and submission. She only displayed bad behavior in the presence of other animals, going all snarling, slobbering psycho on anything else with four legs and a tail. It was quite a transformation. For Abbie and Tucker—and most Australian shepherds, German shepherds, and other working dogs—plenty of exercise and ensuring they felt like they had a purpose in the family structure helped tremendously to mitigate their tendency toward being overwound.

Combining these aforementioned elements of irrational and irascible were my greyhounds in the warehouse district of west Berkeley: poor wire-crossed retired racing dogs, whose versions of deranged made them equal parts dangerous, sad, and disgusting. Sometimes I looked into their eyes and it was like no one was home; they acted out of a bizarre, squirrely instinct that involved crapping on each other's heads and snarling at anyone in uniform.

And then there was Felicity, of course, whose bite had been worse than Foxy's bark—or any other dogs', for that matter.

There were countless other dogs and corresponding depraved behaviors (the Mulcher, the thong-eater, or my flatulent friend Charlie), but Abbie's recent antics took the crazy cake. True, she was only a danger to herself, but, in terms of completely bizarre behavior, she was the winner winner, antidepressant dinner.

It all made me wonder if I was perversely attracted to clients—or they to me—with emotionally and/or mentally challenged pets. Or maybe this was par for the course with all animals. It was conceivable that each came with at least one problematic aberrance. Humans certainly seemed to be built that way. And most had more than just one glitch in the system—I certainly did.

Most troubling for me was the slow realization that many of the dogs I'd encountered over the past year were proving to be as opaque and unpredictable as I'd long found humans to be. When reading a dog's emotional state or anticipating their likely reaction to a stimulus, I was guessing wrong as often as I guessed right. When dogs are your full-time job, with a side of cat or fish or bird thrown in for good measure, being wrong 50 percent of the time was bad news. The truth was, they could easily turn on each other—or me—with little advance warning or justification that I could intuit. It wasn't always, or only, fear-based. Or was it? More beneficial than becoming an expert in animal behavior, it seemed, would be specializing in animal misbehavior.

Without that first beautiful and perilous walk with Abbie and Tucker, I went straight to the bottom-of-the-hill clients. First was Morgan, who occupied a cozy little cottage with her owners in a neighborhood of similarly small and charming starter homes. The owners were young, maybe early thirties, and Morgan was as yet their only child. They left *Animal Planet* on the TV for her while they were away and insisted I sing to her to get her to poop. They swore it worked and didn't want to run the risk of her getting backed up, so I obliged. Though I sort of cheated, taking her in the backyard before our walk and singing to her there, for some measure of privacy. I didn't exactly want to be standing on the sidewalk in full view of God and everyone saying, "Moooorgan, make a poo poo," in a high falsetto. It was my poor luck that their neighbor was a big gardener, often in his yard on the other side of the fence and serving as my frequent and unwitting audience.

Next, I continued on up the road to Spence and Doodle's. Doodle could not be cajoled, bribed, trained, or physically dragged to move any faster than a leisurely amble. If I went too fast for his sensibilities, he'd just lie down right there, whether we were in the middle of a crosswalk, trying to avoid the reach of a sprinkler system, or giving wide berth to a car that had been circling the block and slowing as it passed me and the dogs. As a result, we spent much of our time together in their backyard. Their owners were the kind that treated their yard as a big grassy bathroom, never cleaning up after the dogs, so I had to be constantly vigilant not to step in one of the moldering mines that freckled the lawn.

There was one phone call I always took with pleasure, which never required that I put out a fire, rearrange my schedule, lie about my health, or elide my financial woes. With my sister, I could be completely candid. There wasn't a single topic I had to avoid or sugarcoat. We could openly discuss, for example, how bad my skin had gotten since I started dog walking. I could tell her that, when Annie took me to the neighborhood nail salon for my birthday, the technician said, "What happened to your face? You used to be so pretty."

On the phone, she and I contemplated whether the sudden and intense outbreak of painful cystic acne could be a result of my weight loss since I'd started walking dogs. Was it stress? Or because I was sweating more? She signed me up for Proactiv and paid for the shipments, rejoicing with me when my face started to revert to its usual ruddy but relatively smooth complexion.

She kept me on birth control, sending me free samples from the Atlanta infertility clinic where she worked. The pharmaceutical companies kept them well stocked. I took the birth control less for its contraceptive qualities than because it kept my otherwise painful and erratic cycles regular and comparatively symptom free. If only she had access to free samples of Celexa or Prozac, I'd be all set.

She didn't judge me when I confessed that I was using MySpace to search for eligible bachelors in the area because I didn't know how else to meet guys. She encouraged me to join a local group of newcomers to the area. This was a Yahoo! Group where the members posted get-togethers: bowling in Albany, karaoke night in downtown Berkeley, pizza-making parties at someone's house, or pool parties at a complex in Pleasant Hill. I'd never been a joiner in my life, but I was getting desperate for friends, for companionship beyond Ian's nocturnal occupation of our living room.

When I met a guy there who made me laugh, but to whom I wasn't especially attracted, my sister suggested I just take the plunge and make out with him. It could be great! I didn't take her advice, but I appreciated her enthusiasm.

Her calls usually came while I was out with the American bulldog. Ace belonged to a fireman, and his walk fell on my way back toward my own neighborhood from the Richmond dogs. He took the most enormous dumps—the dog, not the fireman—and I'd have to put my sister on hold, balancing my scuffed flip phone while scooping the steaming pile in one quick movement.

"Yuck! Done," I'd say as I came back on the line, and we'd continue on with the truth telling.

For her, I broke my cardinal rule of not talking on the phone while walking a dog. Better that we should talk while I was petting a dog in a shit-speckled backyard, and not out on the street walking one, but she called during her lunch break. Her window of availability was very specific, and I didn't want to miss the opportunity to shoot the breeze with her. Ace wasn't the best dog to break the no-talking-on-the-phone rule with, either. He was incredibly strong and certainly possessed the strength to drag me down the street should he so desire. Thankfully, though, he also responded well to voice commands and didn't seem to be aware of his own incredible strength.

My dear sister loved hearing stories from the front lines of pet care, sometimes asking me to retell her favorites. She cackled through the receiver as I related the latest association meeting in which we observed a Reiki demonstration on a kennel dog, a pit mix named Peanut.

"I don't think he was relaxed; I think he was paralyzed by fear. Twenty-five dog walkers watching him have his life force energy get moved around on a folding table? I dunno, maybe I'm high and it totally works. I'd probably benefit from some Reiki right about now."

Her laughter buoyed my spirits and gave me hope that this was amusing, and interesting, and worthwhile, what I was doing out here, a country away from her, the rest of my family, college friends, and familiar southern sensibilities.

While interesting in a puzzling and perplexing way, canine suicide attempts weren't something I'd normally chuckle over. Nor were my precarious finances, slipping serotonin levels, pizza face, shiftless roommate, and overarching feeling of isolation. My sister helped me immeasurably in gaining a little bit of perspective on my problems, and, after a conversation with her, there was little that felt so hard or so bad after all.

Though my family was far away from me, and I kept my parents only partially informed about the reality of my situation, I knew they all three had my back no matter how dire the times, how shitty the situation, how desperate the measures.

Journal entry: Friday, 6:00 PM

TGIF. Today has been an unmitigated disaster. I dared to snooze for an extra thirty minutes, so my first two dogs had accidents in the house. I think Louie intentionally ran in circles, mid-crap, just to make as big a mess as he could. (Lucky thing about oriental rugs is all the fleur-de-lis patterning . . . I challenge anyone to find the stains!) The rest of the day went downhill from there—got locked out, the dogs ate something off the counter and barfed all over the bathroom rug—and now I have an interview with that 150-pound mastiff client. I'd really like nothing more than to get out of these muddy, wet clothes, crack a giant bottle of wine, and fall asleep in the tub.

CHAPTER EIGHT

Rain or Shine

....

The contracts I had with my clients stated that the dogs go out rain or shine. Depending on the dog, though, their need to piss and shit did not necessarily mean they ventured out into the weather willingly. This was true of the dogs I was scheduled to visit first that day: the thin-skinned and neurotic greyhounds.

Flannel and Salvador lived in a three-story industrial loft in an up-and-coming, semi-industrial business and shopping district. As the hired help, I was allowed to park behind the building in the *No Parking—by threat of death* parking spot reserved for the tenants. When I stepped out of the driver's side, I had to squeeze myself and my giant fanny pack between my car and their motorcycle, parked alongside the back fence. I wasn't sure whether it was Matt or Darlene that biked, but suspected it was a shared passion, like adopting retired racing greyhounds.

To get into the loft, I first had to get through the tall wrought iron gate that opened into the small courtyard. The "dog room" looked out onto the courtyard, and I always took the utmost care not to wake the slumbering greyhounds when I came to the front door. The moment they sensed my presence, they went into a wild frenzy. I'd been cracked on the jaw by Salvador's bony head before, and, ever since then, I was very careful to keep the dogs as calm as possible.

On this particular visit, the dogs must have heard my car or the clang of the gate. Both dogs were jumping and yipping madly before I even unlocked the plate glass front door. Salvador was the larger of the two, white with mottled gray-brown markings on his body. Flannel was more petite, but just as excitable. Her brindle coat stretched alarmingly over her frame. Her skin was so thin, I sometimes thought I could see her organs, pulsing and shuddering as they worked.

Flannel and Salvador's owners were a rare case, in that I preferred the humans to their dogs. Matt and Darlene always paid on time, and in contrast to their hyperactive dogs, they were measured and approachable people. But, cool as they seemed, I never got over the squalor that their dogs lived in. Even though they got a whole room to themselves, which was separated from the rest of the sprawling interior by a wooden baby gate, the polished concrete floor was filthy, and their food and water dishes always looked as though they hadn't been washed in ages no matter how recently I may have given them a good scrubbing. Wet food congealed around the rim of the stainless steel, and the remnants of the dry kibble were scattered all across the floor. More than once I'd arrived to find that an army of ants had discovered the feast. When the dogs drank, they splashed as much water on the floor as they consumed, leaving a lake of water dotted with swollen soggy kibble, which they tracked all over the room.

I couldn't help myself and often grabbed a roll of paper towels from the kitchen and did a quick wipe down to get the worst of it up before we headed out on our walk. Usually, I stopped short of washing their dishes, too, as that felt like an obvious judgment and might have been insulting to Matt and Darlene. But I was always sorely tempted. The ants gave me a welcome excuse to wash down the whole area, all bowls included. I don't know if it made the dogs feel better, but it eased my mind.

Each dog had a handmade cushion sewed by Darlene, which they spent much of their days lounging upon. The large island in the center of their room was heaped with mounds of mail and cryptic knickknacks. There was a six-pack of dusty, empty root beer bottles, a pile of costume jewelry, and a Rubbermaid container filled to overflowing with doll clothes. At the edge of it all sat Dog-Walker Barbie. This was the conduit through which I communicated with the owners, though it was usually with Matt.

The Barbie had shown up on the table months ago, blond hair braided, dressed in a green cargo skirt and rain boots, all strikingly similar to an outfit I'd wear to walk the dogs. I wasn't quite sure whether to be amused or offended, whether this was a friendly joke or a creepy slight. Tucked within the mini messenger bag was a note that read, *Flannel doesn't like to go out in the rain. If you can't convince her to go out on a walk, take them into the backyard instead. –Matt.*

In most cases, it wasn't the dogs that minded the rain, but their owners. I'd been visiting a Jack Russell named Kimchee, whose humans were about to have their first baby. The husband was very explicit that he wanted his dog thoroughly rubbed down and blow-dried after a walk in the rain, his paws washed of all mud, his sweater hung up to dry in the laundry room. Their house was decorated in shades of eggshell and ecru, so I could see where dirty paw prints might mar the aesthetic. For Kimchee's part, he enthusiastically romped through the rain, getting as filthy as he could without actually rolling around in the swampiest puddles. *Best of luck with that baby,* I thought as I blew Kimchee dry, the two of us sequestered to the front doormat until all evidence that he'd been outside—that he was an animal, after all—was expunged. I wondered if Kimchee's dad knew he was getting a dog when they brought home the Jack Russell, or if he thought he was bringing home an extra-animated throw pillow.

Flannel's aversion to getting wet was just one of an encyclopedic list of quirks I'd noted about these greyhounds in their crowded file. Foremost on that sheet of idiosyncrasies was her unusually placed urethra, which caused her to pee straight backward instead of down onto the ground. More times than I care to recall, I'd been following too closely and had my Chaco-shod foot sprinkled with hot urine. She also pooped between three and four times on a walk, and it was rarely, if ever, solid. On Salvador's list of bad habits was his proclivity to walk with his snout permanently placed between Flannel's hind legs, which had resulted more than once in Flannel pooping directly on the crown of his head.

I had long ago given up on getting the dogs to walk in the rain. Flannel especially could only be coaxed out if I held an umbrella over the dogs. Juggling both leashes and a polo umbrella large enough to accommodate two fully-grown greyhounds in their doggy ponchos left me out in the rain and without much leverage to manage them. And managing them was critical. The dogs lunged at anyone in uniform, subjecting the mailman, UPS delivery people, even the poor kitchen staff at the nearby bakery taking their cigarette breaks, to their sudden, bug-eyed snarls. I could rarely predict what new and obscure moving object might trigger them. Skateboards, strollers, bicycles, rollerblades, and the dollies used by delivery people had all set them off at some point.

Compounding all of this was their aggression toward other dogs. Luckily, the shopping district was more commonly populated by well-dressed professionals, tourists, or wives and mothers spending their day out doing their vestigial gathering duties. But other dogs certainly weren't unprecedented.

Early in my walking relationship with the twitchy beasts, we were returning home past the ceramics gallery two doors down from the loft. In what felt like an instant, Flannel had slipped backward out of her collar and escaped my rein on her. Lest I ever forget that she was once a racing dog, her speed was unmatched. She charged through the open door of the gallery, her target the proprietor's geriatric toy poodle. Salvador and I charged after her, bringing the total of dogs in the small display room up to three.

"Flannel, no!" I bellowed. I used one foot to bar Salvador from joining in the mayhem, using my other leg as a wedge between Flannel and her tiny victim. Somehow I managed to slip her collar back over her head and pull her away from the prized centerpiece of the room, a giant raku vase on a pedestal, which she had been savaging the poodle beside.

From the threshold, the dogs behind me strained for more action within. I shakily apologized to the owner, who seemed miraculously unruffled.

"This dog has escaped death more times that I can count. When I found him, he was out on the tracks. Been hit by a train!"

Looking at the shivering handful of kinky cream fur, I found this hard to believe. Yet this death-defying poodle seemed largely unscathed, despite Flannel's best efforts. No blood or broken bones that the owner could detect in her rather cursory examination. I'd been so sure in those frantic moments of separating the dogs that I was looking at a broken neck—and some pretty pricey broken pottery—on my record.

Apologizing again, I retreated with my charges, returning them to their holding area before they could do any more damage. I walked away with a bloodied toe from Flannel's untrimmed nails and the residual chest pains from a minor heart attack, but no lawsuit. My toe was still attached, and my poor heart would surely recover. I would certainly request special greyhound collars for both dogs, too. Since greyhounds' bony, narrow heads are smaller than their necks, the specially designed collar constricts as they pull, and escapes like Flannel's are thus prevented. I cursed myself for not suggesting this to Matt and Darlene sooner. It should have been first on my list when I signed them as clients. That was a bona fide bush-league error. After that, we always walked across the street from the gallery, eliminating any possibility for a rematch between greyhound and immortal poodle.

Because navigating the street scene with the two loopy greyhounds was already a trial without the additional challenges that rain presented, I preferred to just let them loose in the semicovered backyard of the loft. This had initially seemed easier than the full-on walk, but it proved to be a charade all its own.

That day, the rain had slowed to a drizzle, and the dogs willingly ran from their enclosure through the open back door and into the yard. I checked the Barbie satchel for a check before joining the dogs outside. I'd left an invoice the previous day and was counting on the money. I had $2.65 in my account and had written a personal check for gas that morning, banking on the chances that I could deposit this larger check before that one cleared. My finances felt like a Rube Goldberg experiment gone awry. If three clients pay me by Monday, I can pay the electric bill. If only Matt and Darlene pay,

I can at least put gas in the car and buy some milk and toilet paper. If I get gas, I can fill the tires up for free. Otherwise, it costs fifty cents. Or maybe Ian can pay this month's PG&E and I'll cover the next one . . .

Ian had recently secured one of the most coveted jobs on the western seaboard with a tech company in the Peninsula and was being shuttled to and from his cushy, high-paying job on a Wi-Fi enabled corporate bus. He left in the mornings wearing jeans and aviators and returned with a backpack full of free gum, Odwalla smoothies, sodas, and animal crackers. Despite this proliferation of snacks in the house, little else had changed with our living arrangement. Meals were plentiful and completely gratis at his office, so the fridge remained bare but for my paltry offerings. He had money now, but he spent it all on his student loans and accumulated credit card debt, and any leftovers went to beer and courting the ladies that seemed as plentiful as the employee perks. I still had to hound him for his share of rent and utilities.

With Matt and Darlene's always-prompt payment, I was grateful that I could bring my balance back up above $100, however briefly. I put the check in my pocket and headed out to the back of the loft, umbrella in hand just in case the rain picked up again and Flannel required shelter.

Somehow the dogs never collided with the armless marble busts and abstract pieces of angular rusted metal that apparently passed as outdoor art. I had to stand stock still in hopes that the dogs might forget my presence long enough to take a crap. If I spoke or moved around, both dogs were sent into a frenzy, feeding off the other's insanity. I'd tried staying inside while the dogs were out back, but they hurled themselves at the French doors until I either joined them or allowed them to come back inside. So I stood like a statue myself, the misting rain accumulating on the short brim of my rain hood and fogging my glasses. Today the dogs peed right away, and Salvador was pooping within minutes. *Bless it,* I thought; I knew both dogs would poop inside without hesitation.

The week before, I'd arrived to find both of them skidding through a puddle of runny shit. After going through an entire roll of paper towels to clean the floor, I was left with only soapy toilet paper to get the brown stains off Salvador's white coat. I left a terse note with Barbie:

Somebody had an accident today. I washed the floor, but it's still a bit stained.

Much as I worried about them getting it all out in the backyard before I restored them to the front room, I was intensely grateful not to oversee their emissions out on the street in full view of well-dressed businessmen and -women, brushing by the bushes in their fine, dry-clean-only suits that Flannel and Salvador had just finished peeing all over. So many times, the dogs dropped their loose loads right on the sidewalk. I didn't come equipped with a hose, and there wasn't much I could do about the dirty brown stain left behind on the concrete to be stepped around or through by patent pumps or a fine pair of brogues. Nothing was worse than having to pull Salvador from between Flannel's legs as she peed all over his face. In the safety of their backyard, the depravity was well concealed from the world and could remain our nasty little secret.

This loft, as with many of the client's homes, was a singular perk of the visit. I occasionally had extra time in my schedule to explore—rarely touching, only admiring. Matt and Darlene's home was a museum of kitsch. In the kitchen was an antique vending machine with hand pulls and a Plexiglas display window. The original inventory was housed within; there were Boston Baked Beans, beef jerky bites, Cracker Jacks, or a square packet of Planters peanuts. There was an impressive collection of vintage lunch pails, which lined the staircase up to the sleeping mezzanine on the second floor. I didn't recognize any of the characters adorning the rusted lids. In the bedroom, the clothes were displayed on a circular department store rack. On a mannequin torso beside it hung articles of Darlene's fabulous wardrobe. A fedora, a feather-trimmed jacket, scarves of many colors.

On the third floor of the loft, the décor was that of a swinger's bar: a leopard-print chaise lounge next to an overstuffed purple velvet sofa. There was a full bar and an antique turntable. French doors led out to a covered deck that overlooked the bay and San Francisco beyond. Nothing in my job description necessitated my trespass on the third floor, but neither did it prohibit a little peek now and then. Sometimes on a Monday morning, the remnants of a party remained: martini glasses littering the chrome coffee table, record sleeves along the sofa—Rolling Stones, Velvet Underground, Bowie. This was a world I could get used to. Below, the dogs scuttled and scratched behind their gate. As always, I descended to rejoin them.

Since Ian had moved in, he'd dated at least two girls I knew of and had slept with as many or more. Sometimes it was hard for me to tell exactly who he was with on the other side of our thin shared wall. I didn't necessarily envy the temporary nature of these connections, but I did note with a measure of melancholy that he was actually meeting people. He was making connections, period, however casual or meaningless they may have seemed to me.

Inspired to meet anyone my age with remotely similar interests beyond Ian, I'd recently taken to utilizing MySpace's search function. I'd posted plenty of pictures to my profile, and, in an attempt at tongue-in-cheek humor, the song that played when anyone landed on my page was Rod Stewart's "If You Think I'm Sexy." I assumed, probably too confidently, that anyone visiting my page would get the meta-nature of the joke. This was probably too much to ask.

I'd also created a rather less sardonic presence on Friendster, another social media site, which had netted me a robust but ill-advised flirtation with my best friend's brother. He and I had tried to date in college, and I felt pretty sure it would've taken had my friend not been so opposed. My loyalties were clear. Yet here I was, a mere two years after we'd definitively abandoned any possibility of a relationship, rekindling the romance with him from afar. Loneliness seemed to be the culprit and my defense for all manner of bad behaviors.

I'd never guessed I might rely upon a social networking site to meet someone. I fancied myself old-fashioned, and relying upon a photo and self-scripted digital persona felt like a more drastic measure than I might otherwise resort to. It certainly didn't seem like the most reliable means of meeting a potential mate, but I had been largely friendless and entirely boyfriend-less for going on one year, and I considered these desperate times.

Thanks to the advance search fields, I had come across a guy a few years my senior that lived in the area, played the trumpet in a jazz band, wrote nice poetry, looked to be quite tall, and had a great job at an ad agency that happened to be located on the same street on which I walked the foul-mannered greyhounds. We'd yet to meet but had exchanged a few safe messages via the site's private mail function.

I think my lack of socialization was starting to erode my innate sense of normal behaviors and boundaries, as I started showing up at his favorite bar and the restaurant where his band occasionally played impromptu sets, hoping more for an in-person glimpse than an actual encounter. Of course, he was never present at any of these locations when I was, and I resorted to drinking alone, ashamed by my lurking. At a certain point, I must have mentioned to myself that this behavior of seeking him out bordered on stalking. I probably replied to myself that it was better than sitting at home hearing Ian's headboard thudding dully against the wall.

As it turned out, it wasn't at his favorite bar, or at any of the venues where he played music, that I finally saw him. I wasn't dolled up, with good hair or a carefully selected outfit. I didn't have a drink in hand, or my game face on. No, indeed. I was out with Flannel and Salvador. In the midst of navigating Flannel toward the bushes to do her business while trying to keep Salvador's head out of the way, I noticed a handsome, familiar-looking figure on the opposite sidewalk, striding in our direction. I recognized him immediately from his pictures, though he was taller and even better looking in real life.

I started to raise my hand in a knee-jerk greeting when I realized instead of slowing or acknowledging me back with his leash- and dog-free hands, his pace noticeably quickened, and he started to look very intently upon his feet, the shrubbery lining his side of the street, anything but me and the dogs by my side. He passed us at record speed, and I was left feeling mortified, Salvador snuffling at my pocket, and Flannel having painted the nearest holly bush brown.

We stopped chatting online after that, and I actively stayed away from anyplace where I could possibly bump into him. I couldn't risk a chance encounter, since he'd obviously avoided me that day on the street. The dogs weren't to blame for this humiliating missed connection with my Internet mystery man, but I blamed them anyway. When I was with them, that memory had a way of surfacing, unbidden, making my guts bubble with shame. Of course, I harbored no illusions about the glamour factor of my job, but I'd never before considered that other people—say, an eligible bachelor I was interested in—might judge me by my work.

After I brought the rain-damp dogs in from the backyard, they got a thorough toweling down and a dog biscuit each. Their water refreshed and the mess of crumbs from their treats wiped up, I latched the baby gate and took my leave.

It wasn't just the rain or Ian's marked success with money and significant others that had me dispirited and distracted that day. On the way up the road toward my afternoon clients, before the bank, I had another stop to make.

The pawn shop, and my destination, was on San Pablo, past the BevMo! near the big shopping center with the Ross and Payless ShoeSource. I passed it daily on the way to my Richmond clients, which had given me the idea in the first place. I had long been toying with the idea of selling some jewelry to help me make ends meet, and I'd finally reached the point at which it was less an option and more a necessity. I'd been wearing the same cracked galoshes for long enough that, despite my duct-taping job, more rain seeped in and soaked my socks than was repelled. I'd ordered replacements—good ones that I could walk many miles in and that would presumably last for a while—but I kept canceling the order to use the money for something more pressing. I still had plenty of duct tape.

In free fall, it's hard to know where the bottom is exactly. Where on the wide spectrum of failure did pawning my personal possessions fall? I was a far cry from compromising my morals and had yet to consider trading anything more than sentimental gifts for cash. But I was feeling pretty piss-poor about the choices that had resulted in my entering a pawnshop with intent to sell. Jewelry. My jewelry. I couldn't bear to part with the sapphire ring my sister gave me for my twenty-first birthday, or the gold tree of life my mom brought back from Egypt, so I settled on an amethyst ring from my aunt and a pair of diamond studs from my sixteenth birthday.

I almost kept driving, but there happened to be a curb spot right in front of the store. I took this as a sign that this was a good idea and I should see it through. Also, I needed the majority of this check from Flannel and Salvador to cover an overdue cell phone bill, and the gas tank was millimeters from the dreaded E.

The facade was fancy, all green marble and gold lettering. But the interior was completely the opposite. Everything was locked down or behind bars. The guy at the counter stood behind Plexiglas. I had to lean over and talk into a little speaker.

"Hi," I said, "I have some things to sell." He looked at me blankly, waiting for me to produce. "I don't really, uh, know how this is done."

He gestured to the narrow slot beneath the divider and the mouthpiece. I pushed the earrings and the ring through, and he scooped them up in his massive hand. He put on an eyepiece to examine the diamonds, and I worried that maybe they were fake and he'd think I was trying to swindle him. I cast my eyes around and realized that my diamonds looked like Grape-Nuts compared to some of the stuff under the counter. He pushed the earrings back through to me. He'd linked them, the one stud through the other and clasped to keep them from separating, in a way I'd never seen before. So tidy.

"I couldn't give you more than $16 for those; you might as well give them to a kid niece or something." I swallowed hard. I'd never give them away. I loved them too much. Yet I'd been moments from trading them for three-gallons-worth of gas. My eyes were smarting at this betrayal in progress, but I refused to let this guy see me squirm.

Meat Hands pawed my ring next.

"For this I could give you $24. The band's worth more than the stones. Just melt it down . . ." his voice trailed off as he fingered the ring, so tiny between his sausage of a thumb and pointer finger.

I held out my hand, implying that I wanted it back. I pocketed the ring and the earrings both, nervous sweat beading my upper lip and dampening my shirt.

"Sorry," I said. But I wasn't apologizing to him. I'd done myself really wrong the moment I slid my baby diamonds and priceless ring beneath that bulletproof partition.

I pushed through the barred doors and into the spitty rain. I still had a sixteenth of a tank of gas left. The warning light would blink on at any moment, but there was still some juice in there. I also had some change in the console, which would get me a little less than a gallon in case of emergency.

Back in the car, I felt safer, and better now that I was breathing normally. I transferred the earrings and ring from my pocket into the cup holder, then changed my mind. I slid the ring onto my left hand and unlinked the diamonds, pinning the studs into each ear. Fancy dog walker. Dog-Walker Barbie. That was much better.

To: "Dad"

Subject: Turkey time

Daddy-o,

I was hoping you could give me some instructions for my Thanksgiving "feast." Mom said you use an old T-shirt to keep the turkey moist?!? Do parts of the bird taste like armpit?? Maybe the butter masks the flavor of Speed Stick and sweat. I have absolutely no idea what I am doing here, so any advice you can offer will save this bird from ruin.

Lindsey

CHAPTER NINE

No Heroics

....

S urely there are more annoying ways to be woken, but as Simpson furiously kneaded my head with his needlelike claws, I couldn't think of any. I rolled away from him, leaving my neck vulnerable to his unshorn nails. Twisting and turning in my half-asleep stupor, I tried and failed to escape his relentless attentions. He worked his way down my back, and I farted on him through the thick blankets. My dream of serial earthquakes and melting cell phones completely dissolved into the bleak gray light of predawn. I tossed Simpson onto the carpet with weak-fisted force, but he was back on me in no time, redoubling his efforts on my head. This was no outpouring of love on his part; Simpson was hungry. He was always hungry.

I was staying with Simpson, his counterpart Larry, and the dog Leilani for the week while their owners were vacationing some-where tropical. They were clients of my colleagues; like so many of my gigs, the work was done in the capacity of a subcontractor.

This wasn't the first time I'd cared for this trio, so I'd already known that this wouldn't be one of those jobs where the clients directed me to the top-shelf tequila, the hot tub, and their king-sized bed. The mattress was lumpy, the fridge was bare but for a bunch of wilted spinach and a door full of condiments, and the cabinets were stocked with inedible ingredients like flaxseed and protein powder. Plus, it was a good twenty-five minute trek to get to their Alameda house from my north Berkeley apartment. That was if traffic was moving steadily through the MacArthur Maze, which it rarely did.

But I'd always liked these clients—human and animal alike. Sara was always smiling, a woman overwhelmingly in love with Andy, her charming and carefree husband. Sturdily built, tanned, and handsome beneath his mussed curls, Andy ran an extreme adventures outfit—skydiving or scuba, waterskiing or the regular snowy kind.

The pictures of their wedding displayed in the bedroom and tucked in nooks around the house showed him in a Hawaiian shirt and sunglasses, her in a simple sundress, her uncoiffed blond hair blowing in the sea air. They seemed real, young, and in love. I liked being in such close proximity to these happy and wholesome people, trying to absorb some of their stability by osmosis.

Their house, down to the odd but pleasing variety of colors they'd painted their walls, always felt joyful and warm. Their animals, too, radiated the contentment that comes with being well loved. I was well aware that there were worse ways to spend a week—and dramatically more difficult animals with whom to spend it.

I'd miraculously managed to wrangle the upcoming Thanksgiving holiday as a rare day off and was celebrating by hosting at the apartment. I'd cook for Ian, his latest in a long string of ladies, as well as my college roommate who'd recently moved to the

Lindsey Grant

Bay Area, the guy she was sleeping with, and an old ex-boyfriend—
the only one I'd ever managed to remain friends with. It was going
to be like the Island of Misfit Toys, but I was bound and determined
to make it a memorable meal.

This would be my first Turkey Day away from home as an
adult, as well as my first time hosting, and the money from these
overnights would certainly help offset the grocery bill for the feast.
I'd been doing some research into the best (read: most affordable)
turkeys, and I thought I could keep the bird to $25 or less, especially
if I went to Grocery Outlet. The guests didn't need to know where I
bought it from. No organic, well-adjusted, Heritage Tom with good
genes and high self-esteem. We'd likely eat a Butterball bird, much
abused in his lifetime, because that's what I could afford.

Even taking into account my appreciation for this happy
house and its occupants, and the utility of the money I was making,
I didn't feel that these pros excused or entirely offset Simpson's early
morning wake-up calls. I was a girl who needed her eight hours.
Unable to imagine his exertions on my neck and back as a form
of acupuncture, I tossed away the comforter, momentarily burying him in bedding. After pulling on my favorite sweatpants, the
heather gray kind with elastic around the ankle circa the Gap in
1989, I climbed over the baby gate that partitioned the kitchen from
the rest of the house.

Leilani, the German shepherd mix, was stirring in her bed
tucked in the back corner of the breakfast nook. The Hansens had
found her years before while on vacation, falling instantly in love
with the stray pup. After very little deliberation, they bought her a
one-way plane ticket on their flight, direct to San Francisco.

Larry the cat was Sara's before she ever met Andy, and he became
equally theirs when they'd married eight years prior. The way Sara

told it, Larry was there before Andy, and had been there for the duration of their life together. *He'll probably be here after Andy, too,* she'd joked. This didn't necessarily give Larry preference over Leilani and Simpson—or Andy, for that matter—but certainly seniority.

Larry suffered from hyperthyroidism, leaving him extremely lean and in need of medication twice a day. In the past, mixing that medication with some wet food had been the only contact I'd have with him in a day. He was an exclusively outdoor cat, only venturing so far indoors as the mudroom for meals and cat naps atop the washer in his round furry cat bed. It was a rare occasion indeed that he joined Simpson and Leilani in the kitchen.

After he ate his doctored meals, he was off again to lead his mysterious cat life. Though he was a scrappy tomcat, a prowler and a hunter, his pure white coat and subdued demeanor lent him an air of stoicism utterly absent in Simpson.

Named after the famous animated family, Simpson was the younger of the cats and was about as different from Larry as he could be. In perfect counterpoint to Larry's uniformly snowy coat, Simpson was jet black with a white bib and boots on all four paws. He was comically obese, and next to the skeletal Larry, he looked even more grotesquely overfed. Where Larry was the quintessential feline—aloof, independent, and completely indifferent to humans and animals alike—Simpson was so desirous of constant affection (and constant feeding) that he made himself a pest.

Fat as he was, Simpson was able to leap from the floor to the kitchen island to the countertop faster than I could swat him down. Whatever I was preparing, be it dinner, a cup of coffee, or Larry's medication cocktail, Simpson would be on it in a flash. So his food came first, always, to occupy him and keep him out of my face. He ate atop the fridge, a vantage point that allowed him to see who else was eating what, where, and when.

Leilani was far more relaxed about her morning meal. I scooped a cup of kibble out of the bin in the cabinet, and the dog was soon settled in front of her breakfast. I intentionally left Larry for last, since his morning regimen was far more involved for the both of us.

Larry had been sick for a while—for at least as long as I'd been pet-sitting him and his housemates. The only changes in his care over that year or so had been how much medication he got and how frequently, and then later, in what he was eating and in what quantities. But any differences in Larry himself had been small and slow, and the Hansens never seemed overly concerned.

When I arrived for this visit, they had left the usual note of greeting on the butcher block in the kitchen. *Lately he is only interested in eating baby food, but he eats it with gusto!* The last of their instructions read, *If he starts to go downhill, no heroics, okay?*

I'd been prepared for Larry's condition to have declined and expected I'd need to watch him closely throughout the week, keeping him comfortable and making sure he got his meds down with the new liquid diet. But that sentence, tacked on to the end of a hastily scribbled note, gave me great pause. I wasn't aware when I'd signed on for this week that Sara and Andy might be outsourcing his end days to me. Once I actually saw Larry, I was struck with the certainty that he would not make it through the week.

They had grossly downplayed the fragility of Larry's state. He'd essentially stopped eating. When I fed him the first night, the most he took in was some juice from the canned tuna I'd put in his chipped blue dish on top of the dryer in the mudroom. I had to force-feed him his medication, which left him understandably pissed at me. Soon after I'd emptied the syringe between his jaws, he regurgitated the meds mixed with gastric juices in foamy dribbles, his emaciated body convulsing.

The worst part, though, was the open and frankly horrific-looking sore on his side, pink and deep purple where it festered. The hair had fallen away, and he worried the exposed area incessantly so that his whole right side was damp and cowlicked. His spine and tail were even bonier than before, a wasted frame under loosely hung fur. His eyes still retained such majesty, though—gray blue and perfectly cat-eyed.

That first morning before I left the house, I got down on face level with him.

"You listen to me, Lawrence. Your parents will be home in six days, and they will be so happy to see you. You hear me?" I kissed him on his head, which had the same velvety texture as it always had. I hummed a bar of The Who's "You Better You Bet," too, because it was the only song I could think of. Then I locked the mudroom door behind me, trying with everything in me to ignore the hot, bilious anxiety rising in my throat.

My time with Larry, Simpson, and Leilani coincided with another cat client in Alameda that requested daily visits for the week. Though the owners had another residence nearer to their workplace, they kept their cats at this second, sparsely furnished apartment by the beach. The street-facing one-bedroom unit was almost entirely given over to the cats, save for an extremely sexy candy-apple-red espresso machine that gleamed from its place on the counter in the efficiency kitchen.

The cats they kept at this apartment were all feral. I rarely saw them but for a tail poking out from behind the curtains in the living room, or glowing eyes under the bed. As I was required to be there for thirty minutes, I took my time changing the water and portioning out their food, trying to occupy as many minutes as I could laying eyes on each of the five cats who lived there. Conveniently, that took up most of the half-hour visit, checking all of the closets and cabinets that were left ajar for the cats to wander freely through or hide within.

If any time remained, I'd stand in the kitchen admiring the machine. Next to the shabby, cat-scratched furniture, amidst the dusty clumps of fur that blew through the drab and dirty apartment, that machine was even more spectacular.

Those half hours spent in the company of the wild, reclusive cats were a welcome change from my nights across the island. Mostly because the cats—antisocial, averse to human touch, and seemingly divorced from any need for companionship beyond that which their fellow cats provided—were so refreshingly different from Simpson's campaign for food and attention and Larry's hospice care. I felt certain that these cats, were they to escape the confines of this carpeted cage, would do just fine on their own.

When I arrived back at the house in the evening, Larry was waiting on the porch banister, looking regal and wise as he tracked my movements, slowly blinking at me as I passed him on my way to the side door.

"*Buenas noches*, Larry! Why don't you come on inside for some loving?" I called to him as I entered the mudroom. I didn't think he was fooled; by loving, I meant medicating.

Leilani heard me coming, and she exploded through the dog door from the backyard, dashing excitedly from the mudroom and into the kitchen behind me. Of course, Simpson was waiting there, meowling with a fury like I hadn't fed him for days. He seemed unaware of, or unconcerned by, his comrade's rapid decline. His emergency, as ever, was getting fed. I quickly dumped a handful of cat food in Simpson's dish before leashing Leilani to go out for her walk.

As starved as she was for some individual attention, I gave her a perfunctory walk. I was preoccupied with the certainty that Sara and Andy had already said their goodbyes to Larry. With that final instruction to me, *no heroics*, they'd reconciled themselves to Larry's imminent departure from their lives.

I was racking my brain as to what I could do for him beyond just sitting by and watching him gamely carry on, however feebly. I wanted to make a *Save Larry* sign or carry him with me on my daily dog walks in a custom-made cat papoose, just so he wouldn't be alone in his last days. He probably didn't mind the solitude; maybe he even preferred it. Like the ferals I was caring for, he seemed to eschew our human fussing and fawning, preferring instead to simply carry on with the business of being a cat. But I minded.

I was all but oblivious to the streetlights coming to life along the residential streets. I instinctively stopped at the stop signs and crossed at the lights, but I jumped a foot in the air when an overwound poodle lunged at us, barking through the wrought iron bars of its fence.

Larry turned away from the Gerber beef stew I offered him that night, dignified in his demurral. I understood his refusal of food as his instinctual way of showing that he was finished trying to survive. Every time I rubbed his gums with the pudding-like substance, I felt like I was betraying this knowledge.

Biscuit was this way at the end; that's how my parents knew she was so sick. One week in early spring, she stopped touching her Alpo. After a day, rats came out of the woods and ate it all up for her. Apparently the vets knew quickly, just from palpating her stomach and organs, that she was riddled with tumors. My parents brought her home for me and my sister to see her again, though they knew it couldn't be for long. She still didn't eat; she just lay in her doghouse on her blanket, watching me come and go with her sweet dark eyes and licking my hand appreciatively as I diligently refreshed her rat-chewed food and untouched water in hopes she might change her mind and chance a nibble.

Once I accepted that Larry's final demise was not a matter of if, but when, I settled into the routine of devoting my evenings to making him as comfortable as he could be.

Before I subjected him to the disgrace of the 1 cc of force-fed meds, he sat for oh-so-gentle strokes. I had to caress his thinning coat just so, or he shrank away like I was hurting him. If I got it right, though, he purred like he was a young cat again, angling his head up, then all the way down, and then twisting around, showing me all the pleasurable spots that remained.

Larry was notorious for spraying in the house, so when—if ever—he came inside, he was allowed only as far as the kitchen. I put his cat bed on the breakfast-nook chair next to my own, and he and I spent those dark autumnal hours together. He slept while I watched syndicated episodes of *Sex and the City* and planned the Thanksgiving menu.

I'd been studying up on my dad's methods for cooking the turkey, which included a broth-soaked undershirt or otherwise expendable-but-clean rag to cover the bird for the first few hours, among other unconventional but essential steps. It was easy enough to ignore all his instructions for making the giblet gravy, as I would not be handling or eating any of the turkey neck. I also requested my mom's stuffing recipe and my sister's famous green bean casserole made with fresh beans French-cut by hand. Even with the preplanning, though, I suspected I'd make frequent phone calls to the whole family throughout the holiday for last-minute directions and clarification.

Before I retired with Simpson to the corner bedroom, I tucked Larry into his deep round sheepskin bed and went through the ritual stroking with him. I wondered if my hands on his neck, back, and rump, which he angled in ecstasy, gave him the only pleasure there really was for him anymore.

Meanwhile, at the cat pad, the ferals seemed to be adjusting to my presence. They actually walked across the living room while I was there in the apartment, instead of keeping to the closets and cabinets or under the bed until I left. I didn't have much, or any, prior experience with wild cats, so this was an education. I was surprised at how beautiful, how dignified each cat was. I guess I'd thought that ferals were crazed beasts with an electrocuted-looking tail and razor-sharp claws, like the cats depicted on Halloween decorations. What I was realizing, though, is that these animals are not defined by their relationships to us—or lack thereof. They are defined by their cat-ness. Which is, after all, what they are. These were not pets, just cats. For the first time I understood that the one could exist apart from the other. Perhaps that was the pleasure that those particular owners experienced: observation of, not necessarily interaction with, the animals they shared their home with.

My last night at the Hansens' arrived, and Larry was nowhere to be found. I busied myself with the dog and cat who I'd all but ignored all week. On our walk, Leilani soaked up my attentions like a neglected child would, poor girl. When I fed Simpson, I pet him until he bit me in irritation. It seemed he and I just couldn't see eye-to-eye; with him, it was either never enough or way too much.

Just as I was about to throw on my coat and canvass the neighborhood for Larry, I heard his quiet call coming from the mudroom. I was so relieved to see his face that I nearly missed his weeping wound.

There was blood and fluid soaking the fur around the sore, which looked angrier than ever. He refused food and water with such polite defiance that I spared him the medication altogether. I doubted very seriously that it would do much except make him cross with me. Instead, I put his bed on the chair next to me and let him be quiet in my company, together in the warmth of the kitchen.

While he rested, I made a baked potato and turned on the television. When I turned my back to him, though, he disappeared. It didn't take me long to find him curled near the pillows in the master bedroom, as close to the smell and feeling of his owners as he could be. In all the times I'd ever stayed at the Hansens', and certainly within the span of that week, Larry had never jumped the gate that partitioned the kitchen from the other rooms. He knew his boundaries and hadn't ever trespassed into the depths of the house.

Before I went to bed myself, I restored Larry to his usual spot in the mudroom. I kissed his soft head goodnight and tried to convey some measure of comfort and love with my strokes, carefully avoiding his sickening sore. I wanted to bring him with me into the house, consequences be damned. But I left him there in his bed on the dryer and locked the door to the house.

On our last night with Biscuit, I had a fight with my parents about sleeping in the basement. I don't remember if they even conceded in the end. I just pulled the quilt off my bed and curled up next to Biscuit in the carpeted area at the base of the steps to the garage. We slept that way, her with her ratty, torn blanket and me wrapped in my own. Or she slept. I just tried to absorb the feel of her, the distinct corn-and-processed-cheese aroma of her paws that always evoked Doritos for me, and the gentle rise and fall of her belly—a place I'd rested my head so many times before. I pet her until my shoulder went numb. I must've slept at some point because in the morning I woke to find my parents standing over us. They took her back to the vet then, where they waited while she went to sleep for good.

At some point in the night, I woke up abruptly to the sound of Larry crying. I threw on my sweatpants and climbed over the gate into the darkened kitchen, flipping on lights and unlocking the door to bring him in after all. But he wasn't in his bed. He wasn't on the steps or in the driveway. He wasn't perched on the banister, watching the street with those intelligent eyes.

In the morning, when Simpson threw himself against the locked bedroom door, I got up quickly. The anxiety of the night before had carried over into my dreams and right into the morning. I bundled up again to go out to the mudroom, but Larry's bed was still empty. I put my hand where the slight imprint of his body fit, but the sheepskin lining was cold.

I left a note for Andy and Sara, due to arrive home that afternoon, in the same place they'd left theirs. *Larry took off in the night and I haven't seen him since. I am probably overreacting. He always comes back!* My false bravado was hardly convincing.

Had Sara and Andy been my own clients, and had I been in a position to engage in communication with them as such, I likely would have been far more forthright about my feelings. But as it was, the cordial, newsy note left on the butcher block was the extent of our business-related correspondence. All complaints and non-urgent concerns were directed to my colleagues; they could decide to forward my input on to the client, or not. Yet another tricky nuance of the business that I didn't fully grasp until it unfolded in real time. I had to learn these lessons as I went, and sometimes they proved far more painful than I could ever have anticipated.

I packed all of my things and made the bed, vacuumed the

floor, and scooped out the litter box. I did the dishes and checked the bathroom and bedside tables for any forgotten jewelry or hair ties. I tossed my stuff into the car, still looking for the shape of Larry by the porch. From a distance, I mistook the white of a grocery sack beneath my car for his shadow.

After three agonizing days of waiting for news on Larry, my colleagues passed along an email from Sara. I'd just returned from the big pre-Thanksgiving grocery-shopping trip, taking my mom's advice to go late on a Tuesday to avoid the day-before crowds and still get fresh produce with which to cook.

Sara had attached a photograph of Larry in his younger days, sitting atop the Hansens' back fence. With it was a brief explanation of how the neighbors had found him, alive but just barely, lying in their backyard. They'd always kept food and water out for Larry, and they suspected he was making one last journey to his spot when he had a heart attack. Andy and Sara did not see him alive, but they buried his remains in their own backyard beneath the lemon tree.

It was the final paragraph that stopped me short. In an effort to lift Sara's spirits, Andy contacted his friend who ran a cat rescue. They'd already adopted a three-month-old named Rhoda. *Can't wait for you to meet her. Simpson and Leilani are already warming up to her,* she wrote in closing. It seemed, then, that they were moving right on, and at lightning speed.

For my part, I would have to figure out a way to get beyond it. He wasn't mine, after all. But the loss felt acutely personal nevertheless.

After Biscuit's sudden departure from our family, we tried to commemorate her legacy. I had written a very maudlin poem about "the unstoppable, rebel force" of her cancer, and how she was "my big white tissue, absorbing my sadness with her gentle brown eyes," which we framed alongside favorite photos of her—carrying a giant branch around in her mouth as though it were nothing more than a stick; frolicking in a rare Atlanta snowfall; lounging on her blanket in the utilitarian doghouse my dad made out of plywood with a sheet metal roof; sitting next to me on the front porch, the wild white blond hair I had in my youth an impressive match to the creamy shock of fur that forever hung in her eyes; and a top-ten list of her best characteristics. She let rats eat her dog food without objection; she always pooped in the ivy; she loved to roll in grass, but most especially in our neighbors' zoysia. I still had her blue canvas collar, white fur tangled in the buckle, with her rabies tag and our address in case she ever wandered too far afield, which she never did.

I was strangely insulted by the rapidity of Larry's replacement. Did his loss not warrant a grieving period? Or was the loss so great, or so long-anticipated, that the best and only salve was a brand-new kitten to fill the Larry-shaped hole in their home? I didn't have that luxury. I had a Larry-shaped hole, too, but only an uncooked ten-pound turkey to fill it.

His death cast a pall over what was an otherwise pretty service-able Thanksgiving. We had two last-minute additions to the guest list: a guy from the newcomers' group that I'd tried and failed to date, and his colleague-maybe-girlfriend. The potential awkward-ness of that scenario was overshadowed only by events of the night before, in which I'd made a pass at my ex. Out of loneliness, for old time's sake, because we were great friends and I knew it wouldn't screw anything up. As it happened, he was back with his ex, and they'd just moved in together, so he'd remained on the sofa in his sleeping bag instead of sharing my bed.

But no one said a word about the slightly charred casserole and stuffing, which I'd accidentally scorched while attempting to warm it in the broiler. And—due to the unexpected extra mouths—we ran out of turkey before we'd eaten our fill. But everyone politely claimed they'd had plenty. To me, a post-Thanksgiving Friday without ample turkey leftovers was unnatural. A failure. But I kept this criticism to myself, relieved as I was that everybody else seemed satisfied with the fare.

After the table was cleared and the dishes mostly cleaned, my mind was filled with Biscuit, and Larry, and the unshakeable question of whether—and how—he'd be memorialized. And, too, a nagging sense of humiliation and aloneness. My ex, out there asleep on our bony uncomfortable couch, had his ex; my friend left with her boyfriend; Ian was with his girlfriend on the other side of the wall; and then there was the guy I'd tried making out with after all on my sister's advice, and his "friend," who'd walked off together into the evening full on my semi-successful first attempt at a Thanksgiving feast. I had plenty to be grateful for; I knew that. I just wanted— more than anything—to have someone to share it with.

In my one act of defiance—and in secret solidarity with Larry and his memory—I never again took care of Leilani or Simpson or Rhoda. I always came up with an excuse as to why I was unavailable for the dates they requested coverage for, gracious in my apologies and always agreeing—through the proxy of my fellow pet sitters— that I just had to meet Rhoda-the-wonder-cat, who was apparently fast friends with Simpson in a way Larry never had been.

My avoidance allowed Larry to remain in memory as he'd always been. He sat there by the porch as night descended, surveying the world with his steady gaze, a gentleman tomcat ready for some dinner and a petting before he departed again to lead his quiet cat life.

To: "Smitty", "SB", "TBug", "LTat"
Subject: Insta-love

My dearest, best girls,

"I can't believe it. I am in love! I am love with a
handsome, talented duke!" (That's from *Moulin Rouge*,
in case you didn't catch it.) Against all odds and out
of nowhere, I have met someone, and he is one of the
most beautiful creatures I have ever clapped eyes on.
I am outta my league here, so if anyone has a copy of
"How to wrangle a man one trillion times better looking
than you, never letting him know that you want to be
his personal baby-maker," can I borrow it?? If I can
figure out a way to see him again, I will lick his snout in
gratefulness for subverting my recent and truly terrible
opinion of men.

I love him. I love love! La la la.

LDogg

CHAPTER TEN

Bachelors

....

T he dog was Baxter, a mixed breed six ways to Sunday that pur-
portedly included traces of Rottweiler, some variety of shepherd, ditto for terrier, and maybe chow somewhere in there, which
seemed dubious but for the dense-to-frizzy quality of his coat, and
a faint purplish hue to his tongue. He was cute as he could be at his
advanced age and with the myriad end-of-life issues he struggled
with. His breath could kill a dragon. His stomach was concave from
a botched excavation of a foxtail that had traveled through his nostril and wormed its way into his stomach. He was arthritic, his bark
sounded like a smoker's cough, and he was so deaf I'd have to shake
him awake when I arrived at his house. Even the vibration from my
feet on the floorboards or the door shutting behind me did nothing
to alert him to my arrival.

His owner, Drew, lived in a chimney-shaped house set high
above Berkeley. It was mere steps from a maze of densely wooded

hiking and jogging trails where we did our walking, or in Bax's case, shuffling. It was a great house: two stories, perfectly square, timber frame, with a wide view of the bay and city below. Until recently, he'd shared it with a wife and her Tibetan terrier, Matilda, as well as their marmalade Garfield, who was missing his ears and tail from a long-ago run-in with frostbite. The cat was still in residence, but the wife and her dog had fled in the wake of a trial separation.

I met her once, when she was dropping Matilda off on her way out of town. She was blond, tough looking, and rode a scooter with a bin in the back that Matilda occupied. For his part, Drew was white blond, rosy cheeked, round in the belly, and spoke to Baxter in the most awful baby talk. His voice would go all high and squeaky, and he'd say things like, "Who's the cutie-wootiest Baxter boy?" I mean, we all talk to animals—I am as guilty as the next person—but he might've reserved the special voice for private moments alone with Baxter. I couldn't honestly say that I saw the attraction for either of them—Drew or his erstwhile blushing bride. The dog-talking voice alone made my ovaries shrivel into hard little infertile pellets.

In the absence of a wife, Drew eventually took on a roommate, presumably to help with the rent. I can't figure any other reason he'd have welcomed this guy into his tiny house. The new roomie was a full head shorter than me, wore braces, had a bong that stood almost as tall as he did, and had no job that I could readily detect. He was often on the couch, stoned, watching basketball and hollering at the screen when I'd come by for Baxter. Why he couldn't mind the dog and save Drew some money, or himself some rent, was beyond me. Not that I was complaining; five-days-a-week walks were my bread and butter, and I was more than grateful to have Baxter locked into my regular schedule.

When the ex-wife's Tibetan terrier was staying over, which happened occasionally, my normally sedate stroll with Bax was a slightly more fraught undertaking. But it was also slightly more money. Matilda absolutely refused to go on-leash, which made me nervous. Baxter wasn't leashed either, but he could barely move without assistance, so I had little fear that he was going to escape into the woods. Matilda, on the other hand, was fast. When she moved, she seemed to leave those streaky graphics in her wake that you see in cartoons when Road Runner takes off at speed.

On more than one occasion, she leapt from the front porch onto the trail and didn't reappear for fifteen minutes or more, during which time my heart was lodged up near my uvula. To rally her, I had to trill, "Matilda, Matilda, Matilda," three times fast and then vibrate my lips, as though I were giving someone a raspberry, at a high pitch. How Drew and his wife determined that this was the most effective way to summon the dog was beyond me. It certainly wasn't dignified, and it wasn't even consistently effective. But this was the only way to make her come back to me.

When she reappeared, tearing up the trail toward Baxter and me, who always shuffled along at least twenty paces behind, it was no guarantee that she wouldn't take off again, so I just kept repeating the same absurd four-part call to keep her engaged. She liked to jump from the ground up to my chest and into my arms in one impressive leap and lick my lips when I made the raspberry sound.

I was surprised to find that, as humiliating as it was to do this in front of the assorted Cal students and outdoor enthusiasts that passed, it was still preferable to chasing her around until she finally got worn out or I was able to pin her to the ground.

Drew was one of a new crop of bachelors I'd taken on recently. There was no accounting for why so many simultaneously single men were in need of pet care. Or, rather, I understood why the newly single would need some help covering for their fur babies, but I couldn't account for the sudden rash of singletons.

Far worse off than Drew and his stoned Doogie Howser of a roommate was a one-time pet-sitting job I took up near Orinda. The house was enormous, part of new construction that had sprung up along the highway. The owner, like Drew, had recently lost his wife, though it wasn't entirely clear how or why. When I met him for the first time at the front door, it was one of the first things out his mouth, part of his apology for the state of the once-grand house.

"Sorry about the house. My wife isn't here." And hadn't been for some time, from what I gathered by the complete and unmitigated disaster that this man and his two enormous and poorly kept Akitas lived in. Pizza boxes were stacked on the kitchen floor, leaning crazily into the cabinets, dishes so long unwashed that they were actually stuck to the marble countertop. Dusty clumps of dog fur the size of my head littered the living room floor, which was empty save for the dogs and a lone leather couch facing a flat-screen TV that took up most of the far wall. The whole place looked like the end of an empire.

He wanted to pay me up front, and in cash, which was rare. I was used to invoicing at the end of pet-sitting jobs and seeing a check in the mail after thirty or so days. At the eleventh hour, he asked if I could come an extra day; he'd leave the additional cash with my payment by the front door. He was going to a motorcycle show and wanted to extend his stay.

I was grateful for the cash payment and thrilled at the sudden and unexpected extra day of pay. It was funny scooping up that stack of cash, as though I was a woman of ill repute, leaving by way of the garage with my earnings. I'd grown so used to crisp, carefully penned checks; this stack of cash without so much as an envelope was jarring.

With the cash, the owner left his garage-door opener, so that I could take the dogs in and out for their walks that way instead of via the rather ostentatious and very steep staircase leading to his front door. I wondered if this was because the dogs couldn't manage the climb. Even though they'd been described as "young, four or five years old," they moved like ancient and infirm shadows of the dogs they might once have been. Their fur was matted, their breath ungodly, and, like most Akitas I've met, they were pretty ornery. But unlike most Akitas, they didn't seem to have much energy—or any at all. Once I had them on-leash, it took some extreme prodding and wheedling on my part to get them off the leather couch. I gave up pretty quickly trying to connect with them, both because they seemed completely immune to, and even a little annoyed by, my pets and ear-and-butt scratches, but also because I quickly discovered that no amount of scrubbing could remove their distinctive stench from my hands.

The stairs within the house itself were almost as plentiful and steep as the ones outside that led to the front door; the house totaled four stories from basement to attic. There were spiderwebs in the stairwell, housing some of the biggest spiders I have ever seen in my life. I wondered if this guy left them there for the company. Arachnophobic to my very core, it took all of my will to ignore them and focus instead on making it down the steps without hyperventilating. My fear is bad enough that I might have declined the job had I known there'd be mammoth spiders in the mix. The stairs themselves did provide a bit of a distraction from the horrors above—the thick pile of the carpet might have once been a creamy white but had since turned the color of old snow.

The owner presumably rode one of his motorcycles to the show, but he had two more in mint condition sitting in his immaculate garage. Clearly he spent the balance of his time here in the basement, and not in the ruins of the rest of the house.

The dogs shat like horses: while they walked, and with even less warning than a horse might give. No tail-lifting, just wet clods of dog shit hitting the pavement behind them. On our first walk, I kept waiting for the telltale pause-and-hunch that signaled a normal dog emission was on its way. None came, and, after we'd turned and walked home, I was appalled to find that someone had not picked up after their dogs, wondering how I'd missed the big piles at near-regular intervals like ellipses along the sidewalk. Trying to avoid the mess, I fell behind the dogs, and only then did I witness their unique and inconvenient method of crapping. What was more, they just plowed through their poop like it was nothing, leaving brown paw prints in their collective wake. No wonder the carpet in the house was a muddy beige.

Over the long weekend of visits, I became adept at trailing behind the dogs, scooping up the loose leavings as best I could. I couldn't really help the marks left behind on the pavement, illustrating graphically the path we'd taken on our walk like the fecal equivalent of Hansel and Gretel's breadcrumb trail.

I tried to avoid touching the dogs—and anything else in the house—as much as possible, and to keep the damage to the public walkways to a minimum. On the rare occasion that I passed another pedestrian, I pretended I had no idea where all that shit was coming from and tried to look appropriately annoyed and disgusted by the liberally soiled sidewalk.

The client never called me again, and I was relieved. I knew it wasn't anything I'd done wrong, but more likely his inability to remember my name or find my contact information amidst the garbage and fur and dishes that were slowly overtaking his once-enviable house.

I was grateful, too, that he never requested an overnight stay. The money might have swayed me toward a yes, which would have been the wrong answer. I did, however, spend the night at Drew's, just once, for a weekend, which was one too many.

This was before the roommate was there; the second bedroom still stood empty but for a few boxes that I guessed his wife had yet to collect. I had the presence of mind to bring my own linens to drape over Drew's. His mattress sat directly on the floor under the kind of poly-fill comforter you buy in a bag with a matching pair of coarse sheets for your first year of college. His wife must have gotten the bed frame in the separation. I wished I'd just brought a sleeping bag and slept on the floor of the spare room. Marginally less cushioning, to be sure, but the intimacy of staying in Drew's room, sleeping on his mattress, was far more uncomfortable. The space was a study in depression: not his own, but the emotion his surroundings evoked in me.

He had a piece of paper taped to his mirrored closet door titled "Goals for Life," broken down into one-year increments; his underwear was tightly rolled and sorted by color in a bin in the closet, left open for me or anyone else to see. But nothing got my nose out of joint quite as much as the condom—unwrapped and seemingly unused, sitting atop the dresser. I didn't investigate too closely, and instead tried and failed to explain away its presence there. Did he forget he'd unwrapped it and left it there? Why did he unwrap it and put it there in the first place? How could he possibly overlook this before leaving for the weekend? I slept very poorly those two nights, wishing more than I have wished for almost anything else that I had not needed to accept this overnight job.

Amongst these unfortunate encounters with love's disen-franchised, I met someone. My roommate from college had recently moved to San Francisco from Atlanta for a position with HGTV, and we were out in the city celebrating her first day at the new job.

Her new roommate was there, a British guy who seemed to have taken a shine to her in the few days she'd been living in their shared Western Addition walk-up. His friend was to join us at some point, and, in the meantime, I was letting their roommate romance play out while I sipped the cheapest beer the bar had to offer.

It wasn't beer goggles or lowered inhibitions that suggested to me that this friend, who finally showed up, was a catch. He was quiet and unassuming, buying us all a round of drinks with little fanfare and an earnestness that seemed out of place in the packed, noisy bar. He had dimples, and a job, and he asked me questions about my job and the dogs I looked after with genuine interest, which made me blush profusely. No one outside of my family ever asked me about work, and in such detail. Certainly not devastatingly handsome strangers. At the end of the night, he donned my coat, wrapped my black chenille scarf around his neck, and walked out the bar. I had no choice but to follow.

Intrigued as I was, my recent run-ins with sad bachelor clients had me feeling rather more than circumspect about single guys. True, this guy, Patrick, wasn't a divorcée—at least that I knew of. By dint of pure coincidence, he worked at the same tech company that Ian did, but in a different field. There was no evidence to suggest that he was anything but normal, and maybe even exceptional, but neither could I banish the thoughts of sticky dishes and unwrapped condoms and malingering animals that sprung unbidden to my mind when I thought about pursuing him. I resolved to put him out of my mind, even though he lived around the corner from my friend's apartment and I was seeing him regularly as a result.

The troubles with Drew didn't start until months after the roommate moved in. Late one morning, on a day that should have been like any other, I opened the front door to the smell of pot, so

Lindsey Grant

strong it seemed to thicken the air I was breathing. Drew's roommate smoked frequently enough that I wasn't unaccustomed to walking in to the scent of weed, but this was something altogether different.

Baxter was passed out just inside the doorframe of Drew's room, and I wondered if it was from a contact high. No one seemed to be home, but I popped up the stairs for a peek into the living room just to see if the smell was coming from boy wonder and his five-foot bong. There were two giant Rubbermaid bins stacked in the middle of the living room, the kind of containers you keep Christmas or Halloween decorations in. Or that's what I'd use them for. I had a strong suspicion that these were not filled with strings of lights or inflatable zombies. And I was alone in the house with them.

I hustled Baxter out of the house and onto the trail, sucking in the fresh air in an effort to clear my head. My clothes were going to reek, too, if I wasn't careful. I wondered if I should have wiped my fingerprints from the doorknob. Or just left the house altogether and foregone the walk? I looked down at Baxter, who seemed none the worse for his fragrant morning.

My phone rang moments after Baxter and I started walking away from the house. It was Drew, canceling the walk.

"I am already here," I said, opting for truth.

There was a silence on the line.

"Ah . . . Well! I trust you can be discreet, then. About whatever you . . . might've seen."

"Right. No problem."

"All right, then. Well, thanks! Kiss Baxter for me." He said this last bit in his special Baxter voice.

I hung up the phone, wondering if it was weird that I thought an apology was in order. Maybe a, "Don't worry—there aren't going to be cops surrounding the place when you get back from your walk

— 193 —

with Baxter," or "Nobody's going to try to hijack the thousands of dollars' worth of illegal drugs sitting on my living room floor while you're there with my dog." If I'd expected it, I sure didn't get it. And when I returned Baxter, I essentially pushed him through the door and locked up behind me, hightailing it to my car like it was base in a high-stakes game of chase.

I wasn't some kind of puritan when it came to pot; I knew that marijuana was common enough in the area and that plenty of clients, and colleagues, for that matter, regularly partook. This was Berkeley, after all. A colleague of mine had called me one night, trying to ask me if I could cover a walk for her but having a hard time articulating her request. She finally confessed that she was high out of her mind because she had snuck a cookie at a clients' house— they'd been sitting out on a plate in the kitchen—never suspecting it was a "special" cookie. She'd done her best to get through the rest of her day but was kind of freaking the fuck out trying to get all the dogs home from her final group walk.

Neither would I try to pretend that I hadn't smoked my own share of weed. I'd gone to college. But I needed Drew to know that it wasn't cool to assume I was okay with what had happened. I couldn't be in his house, alone, if some shit went down. So, crying inside at the loss of wages, I sent Drew a letter two days later resigning as Baxter's dog walker.

What I received in response to my letter was a sincere apology, a humble request for me to resume my visits, and the promise that this would never, ever happen again. I didn't totally believe him, but I was grateful for the lip service and elated to keep Baxter on as a regular walk. By sending Drew my resignation, I felt I'd done due diligence in preserving my unimpeachable reputation among my peers, and no one would judge me too harshly for taking Drew and his ancient dog back.

For a short while, everything really was okay. Drew commissioned me to take Bax to the self-serve doggy wash down by the bay. I loaded him into a cubicle custom-built for pet baths, shampoo provided, and scrubbed him down, which he seemed to love. He kept looking over his shoulder at me like, "Hey, isn't this fun?" I was almost as wet as he was and smelled just as much like a wet dog, but it *was* fun, like a field trip.

As it turned out, I was extra glad to have that bonding opportunity with Baxter.

I was driving into San Francisco for my first official date with Patrick when I picked up a message from Drew. Baxter needed to go to the vet, and he wondered if I could take him. I couldn't tell from the voice mail whether Bax's need was urgent, or if this was a longer-term "at some point" request, like the bath had been. I was cresting the hill on Geary, about to descend toward Japantown and my date-to-be's apartment beyond. I decided to be safe and wait until I'd parked to call back. The proximity of that neighborhood to both the Fillmore and Pac Heights made finding an open and legal place to park extraordinarily difficult, but patience and luck won out, and I eventually scored a spot just around the corner from Patrick's apartment.

Drew picked up on the first ring.

"Hey, so can you take him?"

"Um . . . I'm in the city right now. What's going on?"

"Okay, well he's been acting weird all day and he needs to go now. Can you come back?"

"Weird how?"

"I dunno, he isn't getting up or moving much, and his eyes look funny."

That sounded like Baxter. I had to guess his usual immobility and the rheumy glaze to his eyes was extra accentuated because that description pretty much covered his usual MO.

"Listen, I am really sorry. I can't make it back this minute, but I could take him in tomorrow when I come in the morning."

"No, it has to be now. I gotta go."

I didn't know in the moment whether to feel bad that I wasn't sacrificing my first date to rush back to the East Bay and Baxter's side, or pissed that Drew was being such a dick. I tried to take a deep breath and simply write off Drew's rudeness on the phone as worry for Bax. Which I hoped very sincerely was unfounded. He was, after all, an old, old man in dog years.

Feeling worried for Baxter, my anxiety altogether supplanting the butterflies I'd had over the very exciting and imminent first date with someone I really liked, I tried to put it out of my mind.

The next morning, I was on my way back over the Bay Bridge from San Francisco because my date with Patrick—the kind and funny and generous and miraculously single and not at all an egotistical womanizer, as I'd feared—had gone that well, when I picked up another message from Drew. Baxter had died in the night, at the vet after all, though he didn't say how Baxter had gotten there. I couldn't help but feel that there was a tone of recrimination in Drew's voice, as though my inability to take him to the vet equaled an unwillingness to save his life. A life which, I was sad to learn, wasn't salvageable.

I continued to walk Matilda for the remainder of the week and then submitted my final invoice to Drew for services rendered, bath included. And he never paid. Despite numerous phone calls and emails, a check for the $150 he owed never appeared in my mailbox. I felt petty worrying over that amount of money, especially in the wake of Baxter's death, which he was surely torn up over. I was, too. But that sum, however small, made an enormous difference in my ability to pay my own bills. Then there was the principle of it, which I tried not to dwell on. I contemplated taking him to small claims court and then quickly abandoned the notion; it wasn't worth the hassle and the bad juju and would probably cost me more than he owed in the first place.

Besides, I was distinctly not litigious. I lived in fear of being sued—or audited, or even scolded—and had recently been threatened with legal action for the first time. She was a potential client; no contract had as yet been signed. She owned two French bulldogs, and I'd met with her to go over what services she was in need of, and to familiarize myself with the dogs I'd potentially be sitting for. It was a tricky job—just pet-sitting, no overnights, but two visits a day, and she lived thirty minutes away. A classic case of the time and fuel canceling out any profit, but a sacrifice I was making more and more just to stay afloat.

Within minutes of our meeting, I knew this was not going to work out. Her specificities for the dogs were so many and so insane—from the temperature of their drinking water to the temperature of the house, the length of time they could be exercised and how vigorously, their body temperature when they finished, and the foods that I fed them, the quantities for each, and in what order—which telegraphed to me in big, flashing, neon letters that somewhere in there I was going to screw something up. Never mind the fact that every room in the house, from the hall to the living room to the kitchen to the dining area, was baby gated to manage which room they remained in when. Depending on the time of day and the activity at hand, she moved them through from one space to the next like steers being herded from pen to pen. The owner attributed the exacting nature of their care to health issues, claiming they had severe asthma and, if they got remotely overheated, they'd keel over and die.

I could already foresee their almost certain death at my hands, and I wanted no part in it. When I told her I couldn't take the job after all, she threatened legal action, as they'd already booked their tickets and were relying upon me to be there for the dogs in their absence. I consulted with my colleagues to see if she had a foot to stand on, terrified that either way, I was going to be facing a lawsuit—this way, for an unfulfilled obligation and the cost of plane tickets purchased, or, if I took the job, for unwittingly overheating her French bulldogs to death. That neither of us had signed any form of care contract entirely exonerated me, and she ended up employing another pet sitter after all. Whoever they were, I worried for them.

I didn't share any of this with Patrick. To my great surprise, he was interested in and impressed by my self-employment and the unusual way I spent my days (and many nights). Were it up to him, I wouldn't do overnights at all, since—devilishly—he declared that if anyone was sleeping with me, it should be him and not a dog. But he respected my work and thought it was extraordinary. (Though he also fully supported my decision to go out on the date with him rather than escorting Baxter to the vet.) I wondered at how he'd been on the market, unclaimed, for so long before he met me, and how I'd been the one to change his status from "single" to "in a relationship." But I also knew better than to stick my head too far into the horse's mouth looking for answers.

I didn't have the heart to say anything that might change his rosy view of my job; it felt too good to be admired. He didn't need to know how much that $150 meant to me, how afraid I was of clients like the French bulldog lady, how easily one mistake or one more lost client could take me from self-employed to unemployed. He'd find out soon enough that this job—fun and kooky as it sounded, and often felt even to me—wasn't all good times, cute pets, and intrepid entrepreneurialism. Sometimes it was just hard.

To: "Mom", "Dad"

Subject: Lindsey Versus the World: Episode 3

Dear Mom and Dad,

YYYYYYah! Whack. Ker-plow! (This is the sound of super-hero me using my good sense and catlike reflexes to battle back against the humiliation of being cowed by the she-bear of life.)

Success #1: I have obtained functioning, legitimate Internet in the apartment and can stop sitting in the car outside the Internet café to use email! (While superheroes are super, they also need technology.)

Success #2: The client whose Bernese mountain dog ate my glasses is going to reimburse me for their replacement after all! Technically they don't have to, since I had "full understanding" that the dog eats anything and everything in sight (underwear, remotes, magazines, napkins), but they decided to be awesome instead of evil.

Success #3: I took y'all's advice and have started looking into grad school. Yesterday, as I was lunching on the samples at Andronico's, I thought about what you said— about realizing my full potential. Maybe if I aim beyond eating free cheese cubes and day-old bread, I'll break the glass ceiling and be able to buy a baguette of my own . . .

Lindsey the Conquerer

CHAPTER ELEVEN

Forty Days

....

I was having a hard time getting out of bed. Having lived in the Atlanta area up until two years prior, I was used to the weather patterns of the southeast. On the edge of the old-growth forest where my family lived, I could sit on the screened-in porch and wait for the coming storm. The sharp earthy smell of ozone permeated the air as the tall trees began to bend and moan in the wind. Those storms were usually wild and exciting, leaving a blanket of broken branches and leaves across the lawn. Within an hour the storm would pass, the thunder rolling on, the cardinals and chickadees and tufted titmice coming out to forage in the new landscape.

In Berkeley, however, I quickly learned not to hope for any thunder and lightning. Here, rain was just rain. Whether it soaked or misted, it tended to come straight down without spectacle. It had been coming down nonstop now for almost a month.

Such uninterrupted rain was unusual in the East Bay. That's

what everyone said, at least. Having lived there for such a relatively short time, I had little historical knowledge of regional weather trends to compare this wet spell to. Initially, the soft spraying sound of the rain in the bottlebrush outside my bedroom window was comforting; I happily burrowed further into my bed, relishing the knowledge that it was wet out there but I was dry and warm inside. Over a month later, I was still waking up to the patter of water on bough and branch, on the dumpster beneath my window, on cement. I knew that I'd be out in it soon enough, just as soaked as everything else in spite of my rain gear.

That morning especially, I could not bring myself to peel back the layers of blankets—flannel sheet, down comforter, and mossy green thrift-store quilt—to put my feet flat on the wood floor and start the day. I dreaded the moment I had to don my perpetually damp rain gear with its earthy fungal smell. I felt pretty sure I was starting to smell that way, too, even when I was showered and wearing warm, dry clothes.

The exposed brick wall in my bedroom gave the space a cavernous feel even in the best of weather, but, with the constant rain, I felt the moisture all the time, no matter how many blankets I slept beneath, or how warm my pajamas were. I'd already tried to move the bed away from that brick wall with its wide window, but there was nowhere else to put it. The bite of the saturated air seeped through the poorly sealed window and onto my head while I slept. When I woke up, the cold was inside the bed as well. Even an inch away from where I lay, the sheets were chilly. What a waste of bed, not to be able to spread out on the wide mattress. What that bed really needed was another warm body, and I was grateful that that was increasingly a prospect these days.

I'd been seeing Patrick more and more, usually at his San Francisco apartment, as he worked long hours and got home late.

The additional commute by BART to Berkeley to see me put him at my place around the time I usually went to sleep. Better to just meet him there on his turf when he got home and have a few alert hours with him before I turned into a pumpkin. I was extra exhausted at the end of these dreary, drenched days. Still, the reward of seeing him at the end of the slog was incomparable, the best and only carrot to lead me out into the never-ending rain once more.

Patrick had been over to my apartment a few times, usually on weekends, when the time-consuming commute was less of an issue. He claimed that he preferred my place to his. He often said dear things like that, having no idea the effect it had on me—that he should like Ian's and my humble, dark little apartment, furnished almost entirely in thrift-store finds or sidewalk salvages. For the first few months that Ian and I lived there, Ian had slept on a mattress we'd collected from outside a house in Hayward, thanks to a post in Craigslist's Free Stuff section. Eventually, the smell and its dubious provenance—and then his new, well-paying job—prompted him to buy a new, unsullied one from Ikea.

For my part, I was still sleeping on the generous queen mattress donated by Annie when I'd moved to Berkeley. It was used, having been stored in their basement for guests to sleep on in the den, but it was wonderfully comfortable and didn't smell one bit.

Growing up, I'd always slept on a twin mattress. At nearly six feet tall by the time I was fifteen, I felt like a hot dog in a bun sleeping in that narrow bed. So here, on this luxuriously large mattress, it pissed me off that I was rooted to the spot, held captive within the small imprint of body heat. Stupid Berkeley apartments with one ancient, fire-hazard radiator, situated nowhere near the bedrooms.

While I lay there trying to muster the will to get up, get up, get up, I ran over my list of dogs for the day. With Baxter so recently out of the picture and no new regularly scheduled walk taking his place, I was on the hunt for additional work. Whether it came from drop-in visits for folks who were traveling or a second job that had nothing to do with animals at all, I was ready to take anything I could find.

I'd had some success picking up some extra pet-sitting gigs, and these thirty-minute drop-in visits for out-of-towners' cats and fish and assorted other low-maintenance companions helped me tremendously in my monthly quest toward meeting the bottom line. But they didn't offer the stability of daily walks, week after week, for clients that needed frequent and ongoing service. Income I could rely upon month after month after month, barring disaster.

I'd gone so far as looking for a second job, a night job, as I wasn't taking on nearly as many overnight clients and could presumably pick up some shifts as a bartender. Learning from my past math mistakes, I'd crunched some numbers and concluded I would make far more in tips from a night of pouring pints than I would sleeping with someone else's pet.

So far, the only place I'd felt confident enough to apply was the bowling alley off 880, a place I'd visited many times with the newcomers' group. Even the bowling alley bar had never called me back, which I admit could've been due to my answer in the "Why do you want this job?" field. I'd said I needed extra money, which is not the most eloquent or inspirational answer. What kind of answer were they expecting, though? Pouring pitchers at a bowling alley wasn't exactly a lifelong career aspiration of mine.

I'd had the prospect of a new five-days-a-week walk dangled in front of me, only to have it bizarrely and unexpectedly disappear days before. That was certainly compounding my inability to throw back the covers and leap enthusiastically into my sodden schedule.

The couple was engaged and had just moved into a low

ranch-style house north of Berkeley. They'd adopted a German shepherd mix, still a puppy, who had boundless energy and would need an hour-long visit Monday through Friday. I was trying to contain my euphoria and relief; such an addition to my roster could be a game changer.

I walked him once without incident, going through the routine training for commands: heeling, sitting, staying, and waiting at corners until I indicated he could continue. I got a call the next day, a Saturday, from the wife-to-be that he'd jumped their back fence and chased a man on a bike around the block, scaring him so badly that he'd called animal control. Animal control, in turn, mandated that the owners build the fence higher, which they couldn't afford to do right away with the impending wedding. So they'd taken him back to the shelter. I was crushed. Crushed, and back to the drawing board.

If all else failed, I was grateful that my parents had leaned on me so hard to complete my grad school applications a few months back. I don't know that they were entirely sold on my choice of creative nonfiction as my master's focus, but I wasn't fluent enough in Spanish to fulfill the required bilingual status for pursuing Comparative Literature, my major in undergrad. I could claim, legitimately, I thought, that writing was a more promising vocation than film, which had been my minor and the likeliest alternative for continued study.

As supportive as they were of my temporary career as a dog walker and animal nanny, graduate school had always been required in their minds. It certainly offered me an alternative to my current conundrum. So long as I got accepted to any one of the highly competitive programs to which I was applying.

I'd sought acceptance to every single university and college that offered a nonfiction option, and that I could realistically attend.

Montana, Colorado, and North Carolina were added to my obvious and original first choice of California. It had initially seemed silly and wasteful to have established residency only to attend a program in another state, but I was soon to learn that getting into a program anywhere at all was a long shot. The supply of aspiring writers—in all genres, not just nonfiction—far overwhelmed the allotted spaces in the available programs around the country. Ultimately, in an ironic twist, the only schools in California that offered an MFA in creative nonfiction were private, so my established residency would do little good in reducing the tuition anyway.

I'd already visited Saint Mary's College, a private Catholic college north of Berkeley where I'd been waitlisted for admission. I sat in on a class and was gratified that they were discussing *Out of Africa*, one of my very favorite books. That might have been an omen of good news to come, but between the warm, dry classroom, and the soft lull of the discussion happening around me, I kept nodding off. It was just so nice to be indoors and not exercising canines or being covered in wet dog fur, and I could hardly keep my eyes open. I didn't even realize I was dozing until the moment my head fell forward, startling me back into wakefulness. The professor carried on as though nothing had happened, but I knew it hadn't gone unnoticed. The tables were situated in a circle, and I was one of only nine people in the room. And then it happened again, despite my best efforts to remain alert and engaged. I tried everything—pinching myself under the desk, taking extensive notes on everything the professor was saying, and even playing a game in which I silently recited every phrase I could think of in the voice and cadence of Meryl Streep saying, "I had a farm in Africa." *I had a snake in college. I ate some cheese this morning. I took a nap in lit class.* It was all to no avail. I could not stay awake.

The specter of grad school rejection and ever-more rain aside, the longer I didn't get out of bed, the more difficulties I'd face in the day ahead. The higher the likelihood that all those dogs I was to scheduled to visit were peeing on the oriental rugs and burnished hardwood floors of their homes, pooping in a closet where it might go undiscovered for a day, or barking until the neighbors called the owner to complain, saying, "No, I haven't seen the dog walker arrive yet." I told myself I was needed. There would be hell to pay if I didn't get my ass out of bed and get going. And still I lay there, loathing the idea of stepping into my ripped galoshes, the mud-caked ski suit that served as my rainwear, and the man-sized blue raincoat that never fully dried out these days and was losing its capacity to repel even the lightest rain.

My visits with each dog were carefully scheduled to prevent accidents in the house and optimize their exercise time, calibrated between the owner's departure and arrival back home. The resulting schedule could be so easily derailed by something as minor as constipation (my own or a dog's) or hitting too many red lights on the way to a client's house. In fact, I'd begun to compress all business and basic needs (going to the bathroom excepted) into my driving time, as dangerous as this could prove. Burritos and subs were the most hazardous foods to eat on the road. After a few near misses, I'd conceded that chasing a runaway black bean or piece of lettuce that had fallen into my lap was not worth a traffic accident. I only made phone calls that required talking—no writing or referring to credit card numbers or calendars or my clipboard with the daily schedules and dog profiles. In the backseat, I had a makeshift pharmacy of tissues, tampons, painkillers, Band-Aids, and an emergency gallon of water. The car was effectively an office on wheels.

My first visit of the day had backyard access through a dog door, so there was no danger of an accident there at least. When I came through the kitchen door, Foxy was lying on her back in the living room with her paws curled toward the ceiling. Her kong chew toy and a roasting pan lay side by side on the kitchen floor, both licked clean.

The living room was packed deep with dusty hardcover books about anatomy or biology. These piles, three or four deep on the end tables and coffee table, seemed to serve as the permanent décor. The front of the house was perpetually cast in a pleasing dusky-hued gloom—in contrast to the bright and organized kitchen—that was inviting rather than oppressive. I had grown up in a house full of books. In my home, the tomes had always been neatly organized on the floor-to-ceiling bookshelves, first by genre and then by author's last name. The smell of the books, organized or not, was the same: musty and strangely warm.

Foxy's shadowy form always drew me across the beige linoleum and into the dim living room, where I knelt beside her on the thick pile of the rug. Upside down, her toothy smile was especially comical. I pressed my face into her downy belly, which was partially obscured by a pink-sleeved baseball shirt, split up the back to fit snugly around her middle, leaving her tummy free for my rubs.

Her shirt is nicer than mine, I thought as I fingered the scissor-cut cotton. I had to admit, I'd subjected Biscuit to the occasional headband, but it never went beyond that. She'd never had outfits. As a child, I barely had outfits myself. I was ever clothed in my sister's hand-me-downs, which were usually from thrift stores anyway.

Foxy was a blue merle Australian shepherd, her fur mottled black, gray, and white. She had pert, floppy ears that felt like mole-skin. I often rubbed my nose against the soft slickness of them. I particularly favored Foxy's smell, which was like warm corn tortillas.

The note on the counter that day read, *Fox chipped her tooth— she ran face-first into a telephone pole on our walk yesterday. Now her smile is that much cuter!* Anyone not indoctrinated in dog expression would interpret Foxy's grin as a menace, but I knew it was a compliment when she showed her gums. This was a typical expression by Australian shepherds—always a means of communicating, whether in greeting or acknowledging they'd been naughty or just a way to connect with you. Foxy's smile was devastatingly cute to begin with, and the chipped front tooth lent that particular charm of a child with dirty hands and scraped knees. It pleased me to no end that her owners didn't intend to have it capped like Tucker's mom had.

Foxy had torn her footpad the previous week and was wearing a white ankle sock over her bandaged paw, the lacy trim lost in the fur of her foreleg. The sock resembled those that I used to wear with my denim jumper—the nicest piece of clothing I owned—to Quaker meetings when I was young. The shirt she wore was supposed to keep her from worrying a hot spot she'd been licking on her shoulder, but she'd been in shirts for over a week, and it didn't seem to be dissuading her at all. Foxy's owners were both med techs, which meshed well with her tendency toward accident and injury. She was absolutely the clumsiest and most oft-injured animal I had ever cared for. It made me adore her all the more.

The endless rain was still coming down, so I wrapped a plastic bag around Foxy's sock. I secured this with the medical tape left on the counter next to the fresh shirt and sock her owners had left out for after the walk. Foxy craned around to lick my face while I worked, the two of us moving in a tight circle through the kitchen; I was trying to tape while Foxy tried to lick. By the time I had bagged, taped, and leashed her for her walk, I was sweating inside my raincoat.

As soon as we hit the sidewalk, Foxy became far too busy to acknowledge me. I might as well have been an inanimate object weighting the end of her leash. The shepherd bounded a few steps ahead, her docked tail unmoving, her nose to the ground. She was at work.

We followed the same route as always, up through the sloped and tree-lined streets of the Albany hills. I loved the daring colors of the humble adobe homes, many with elaborate gardens in front. I was forever ducking low-hanging branches and dodging the roots that busted up the old concrete of the sidewalks. We emerged from the residential cluster at the top of a long commercial street stretching west toward the bay. This was one of few urban dog walks I did, second only in foot traffic to the dicey shopping-district walks with Flannel and Salvador. Walking among businessmen and businesswomen out to lunch or off to the bank made me feel connected to humanity in a way that the neighborhood or trail walks, removed as they were from the bustle of workaday life, could not. Instead of deer or other dogs, I got to observe my own species at work and play.

As it turned out, the commercial drag was a great place to walk in the rain because the storefronts almost all had eaves or awnings deep enough to shield me while Foxy bobbed and wove down the sidewalk, pissing on a fire hydrant or up a tree trunk. Even in the weather, pedestrians rushing around in their business suits with jerking umbrellas would stop and notice this dog, comical-looking with her plastic-wrapped paw and pink baseball tee. It seemed a unanimous consensus that she was cuteness embodied.

Though she had a dog door with access to the fenced backyard, I always felt better if she pooped at least once on our walk. It was— with all dogs—a sign that all is well; the system is up and running. Monitoring the consistency for any aberrances, too, was a great litmus test for overall health.

As she made her ritual tight circle in a sparse tuft of grass in front of the high-end deli, dancing a little before she committed to a squat, I realized that I'd left the doggie bags on the counter. I could see them there in my mind's eye, slowly uncrumpling next to the roll of medical tape. Trying to clean up poop without a bag was always a pain and just plain unpleasant, especially on a crowded thoroughfare with so many well-dressed witnesses.

I'd been in this situation plenty of times, and it called for quick thinking and creativity. Sticks, leaves, receipts, chip bags from the garbage—whatever worked. If I was really lucky, I had a leftover napkin or tissues somewhere on my person. At least on this street, the trash cans were plentiful, and I wouldn't have to carry the bundle for any distance before I could dispose of it.

As Foxy did her business in front of the suited men and women coming out of the store, their roasted salmon and quinoa salad or handmade farfalle packaged in heavy paper bags, I cast about for something, anything, I could use. I needed a piece of one of those nice paper bags.

As I was contemplating using the plastic wrapped around Foxy's paw, I spotted a corner of cardboard under a utility truck a few cars down. With the rain coming down harder now, I only had a few seconds before the poop was too soggy to scoop. I grabbed the cardboard, which was still stiff and pretty dry. While Foxy shook out her butt fur and trotted gaily away, I bent down and swiped at the pile. Success was mine, and I flung the mess into a garbage can with relief. These maneuvers didn't always go so quickly and cleanly.

The trash can was right in front of a pet adoption center, and I realized with a sinking heart that there was a bin of recycled bags there on the wall, available for anyone's use. How I'd missed that detail, I couldn't figure. I grabbed a few for insurance, shoving them into my fanny pack.

Foxy loved to bark and lunge at the little Chihuahua mixes and terriers out for a quick pee break. With Foxy in tow, I could never pause for too long to admire the kittens in the window, piled onto each other like pancakes. "Foxer, come on! No dogs for lunch," I chided as Foxy gave a particularly abrasive yip at a dachshund that came waddling through the bottom half of the green barn door. He was wearing a yellow rain slicker one size or so too small for him, making him look like a little canine Chris Farley.

Once, Foxy had gone after one of these shelter dogs with such ferocity that her leash, going taut, took down a plastic table at the adjoining Mexican restaurant on the corner. But every time, no matter how wound up she became, Foxy would get to the park, only fifteen feet or so beyond, and seem relaxed again, bounding homeward without a trace of aggression.

Foxy's owners left one or two threadbare towels just inside the door for me to dry her with. With the recent weeks of rain and mud, I could've easily made washing my collection of well-soiled towels a second job. I had repurposed any absorbent material I could justifiably destroy, quartered into functional shapes and sizes. This included a terry cloth shower curtain that the dryer had inexplicably burned, a pair of hospital hand towels I got at Goodwill for a quarter each, an old bed sheet ruined by a blood stain, and one of my nicer towels that had been bleached by Ian's many creams and tonics for his psoriasis.

Despite my best efforts to keep a clean stockpile in the back of the car, I was fresh out at the moment. I knew it was inexcusable to be unprepared like this; it was just so hard to wash those towels every evening when I returned home, soggy and soiled myself. I convinced myself that I was allowed to wash me before washing them.

The washer and dryer down in the basement of the apartment building were housed in a cozy, cobwebby room jam-packed with every variety of laundry detergent imaginable. All the bottles and boxes were nearly empty, many so dusty and neglected that they presumably belonged to tenants who had long ago moved out. I'd worked my own economy-sized box of All down to the rock-hard remains in the bottom corner and was not at all above pilfering the dregs of others' easily accessible soaps. I was also a firm believer in the power of hot water to get something mostly clean, soap optional. Each load of laundry cost $2 to wash and $1.50 to dry—meaning the clothes came out hot and damp. I'd been liberally pilfering quarters from the impressive bowl of change, a mix of American and international coins, that sat on Ian's desk.

Wriggling out of my own sodden rain gear, I left my galoshes out on the front stoop. Because they were cracked along the sole from so much wear, my socks were soaked through from the walk. I'd brought an extra pair for just this reason, but, at this rate, I wished I'd brought a few more to carry me through the day.

In the kitchen, off came the plastic and then the sock on Foxy's paw. I would have done well to bag my own feet; her injured paw was completely dry thanks to the dressing. Her shirt came off last, and she shook herself vigorously, seeming to luxuriate in her nakedness. I gave her a good rubdown and kissed her velvety muzzle, irresistible in its proximity to my face. With fresh gauze around the injured paw and a clean sock in place, we were free to do her hair.

In the bathroom off the front office, which was painted lavender and as equally stacked with books as the living room, was a blow-dryer and dog brush. I wondered if this was Foxy's favorite part, feeling the heat of the dryer permeate the layers of her fur down to her skin. I sat on the toilet seat, brushing her out and shaking the dryer over Foxy's tufty butt fur, where the shirt had not protected her from the rain. Her eyes were closed, and she leaned into my knees.

Somehow I didn't mind brushing and drying Foxy post-walk the way I did Kimchee. I could actually imagine that Foxy enjoyed the process and felt good once she was clean, dry, and coiffed. In Kimchee's case, it seemed like he'd get greater pleasure out of tracking mud and grass and all manner of other dirty nature all over his owner's creamy pristine décor. I admit, I would have loved to watch him do it, too.

I recalled the feeling of my sister playing with my hair when we were young—a favorite diversion for her, and something I usually tolerated rather than enjoyed. But there was a distinct pleasure in feeling the air from our ancient pink Conair flowing over my shoulders. I had always had little patience for being fussed over, to my sister's constant dismay. She loved to dress me up in her clothes and curl my hair into ringlets. It was only for the heat from the blowdryer that I stayed put on the toilet seat, towel serving as a cape, when she begged to play hairdresser.

She still lived in Atlanta; the townhouse she and her husband bought was a mere mile from the house where she and I grew up. Our parents still lived there in Atlanta as well, though so many years later, the house little resembled the humble red brick ranch we'd moved into when I was three and she was seven. The pink-tiled bathroom where she decorated me like a paper doll was now remodeled with a sophisticated pedestal sink and olive green walls. We had played bath class in the old tub every Wednesday and Sunday night until she hit puberty. She was the teacher, instructing me on how to properly wash my hair or wind the wooden tugboat so it would reach the opposite end of the tub. Class was canceled only once in all those years, when I got upset with her for getting soap in my eyes and maliciously pooped in the tub.

When I turned off the dryer and unplugged it, Foxy opened her eyes and smiled up at me, baring her snaggletooth. I pulled a dry shirt over her head, manipulating her legs through the powder blue sleeves one at a time. The front of the shirt fit across Foxy's back, boasting *Foxy Lady!* in glittery gray and blue curlicues. At the kitchen door, I pressed my face against her dear head once more and pulled on my heavy jacket.

"See you tomorrow, little girl," I called as I locked the door and stepped back into my tattered boots.

In an effort to patch up the hole in income left by Baxter's passing, I'd taken over a twice-a-week walk for a colleague whose schedule, unlike mine, was full to overflowing. The dog was Princeton, a horrifically bourgeoisie name, but then it would be fair to say that he was born with a silver bone in his mouth. His owners lived in a veritable villa tucked deep in the Berkeley hills, and I was but one service provider on that estate crowded with contractors of all kinds. We were like so many worker bees maintaining a hive.

When I arrived at the parking pad outside the main gate, I could see the gardener's truck was also there, a telltale hose snaking down the curved staircase to the wide front lawn. I could only assume they were in the interior courtyard, as all the rain was more than sufficient to water the outdoor plants.

Upon entering the foyer, I could see that the cleaning ladies were there as well, the rug rolled to the side and a mop bucket in the corner. I skirted the edge of the room to avoid messing up the freshly mopped floor and listened for the click of Princeton's nails on the terra-cotta tiles.

I gave a tentative whistle and a clap, hoping he wasn't shut up in a room with the husband. Extremely young to be semiretired, he conducted what remaining financial business he dabbled in from home. The house was so vast that I rarely saw him and could easily lose the dog as well if he was cloistered behind any one of the massive wooden doors that lined each level.

Princeton descended from the curved staircase before me, his silky shining fur flowing majestically. He was easily the most handsome golden retriever—or dog of any kind—I'd ever had the privilege of tending to. And he was a sweetheart to boot, the perfect pet specimen in every regard. At the base of the stairs, he greeted me with an enthusiastic lick to my hand and a full-body wag.

"It's raining again, buddy. I'm sorry." To the owners' credit, they neither required nor even requested that Princeton wear any kind of rain gear—no waterproof booties, fleecy harness, or puppy poncho—for our wetter walks. He was free to walk in the rain and get just as wet as Mother Nature intended.

Our twice-weekly walk was comprised of a few laps around a big man-made lake at the nearby park. Much of the path was paved, but it could get quite hilly and treacherous in the heavily wooded portions, the soft paths churned to mud and ever-widening puddles. Depending on the intensity of the weather that day, we'd brave it, or else do extra runs of the cleaner concrete sections.

Either way, Princeton was easy. He never resisted, always listened, and seemed utterly content whether he was dry and sun-warmed or soaked to the skin. He was such a classic golden: affable, loveable, and easygoing. It made me feel bad that, of late, I'd been the exact opposite, and a poor match for his happy disposition. Sometimes I just got so dispirited by the relentless rain and the same circular route that I'd just sit down on a bench, Princeton settling by my side without objection, letting me pet him and talk to him about how sick to death I was of being wet. For him, being inside at his owner's feet all day, he was probably happy for the fresh air, whereas I'd have given a lot to spend an hour or so of my workday out of the elements in a warm, carpeted room.

I'd tried so many times in so many different ways to reengage with my daily routines. To celebrate that I got to spend my days with dogs and cats and was not tasked with solving unsolvable problems. That I was my own boss, for all intents and purposes. That I got to move around all day long, getting plenty of exercise outside instead of sitting at a desk. If I couldn't take my days moment-to-moment, embracing the process and the perks of my job, then better to disconnect from it altogether, zoning out until I was finished performing the many repetitive and endlessly wet tasks. Anything seemed better than feeling so dulled by the monotony and isolation of my hours. I chalked up the intensity of my apathy to all the rain—everyone in the area was talking about seasonal affective disorder—but it certainly didn't help that I still wasn't back on my antidepressant. Every month I tried to come up with the extra cash, and every month I fell short of the prescription cost.

It never failed that, upon my return to the palace with Princeton, the cleaning ladies had moved from the foyer to the kitchen, where all of Princeton's brushes and treats were stored. I tried to be unobtrusive as I ducked into the far cupboard for the supplies, taking them into a beautifully furnished den, where Princeton sprawled out on the floor for his grooming. I cleaned his paws, today resorting sheepishly to the use of hot soapy paper towels since my own unwashed towels would probably make him dirtier than he already was. Then I combed out the tangles in his fetlocks and gave the rest of his fur a good brushing.

Before I left, I was careful to leave a note indicating that I'd come. This was a recent request from the owner, as they claimed they never knew when I'd come or whether I'd been there at all. I took slight offense to this, taking the request as a suggestion that I took advantage of their massive house and the many workers crawling the property to skip out on Princeton's constitutionals. *Good walk today, if a little wet!* I wrote. I didn't feel particularly inclined to embellish further; our walks were always exactly the same, and I was without the inspiration to dress it up. They had a perfect dog, and I was burned out. I wondered how a note to that effect would go over.

My next appointment couldn't have been more different from Princeton's pristine and stately environs. If his home was heaven, these dogs were surely living in some version of hell.

One of the dogs was an extremely aged Samoyed with such advanced arthritis and probably hip dysplasia that she literally got stuck where she was and couldn't get up. She had flashes of mobility, but, when she locked up, she couldn't help herself out of it. When I entered the backyard by way of a locked side gate, I found her more often than not lying in a mess of her own waste. I had to hose her undersides down before physically lifting her and moving her to a cleaner, dryer location. This was complicated by the other dog, a three-legged pit mix who was fiercely protective of his companion. He barked viciously at me, bounding around lopsidedly, until I'd cleaned her, moved her, and stepped well away.

Every time I visited these dogs and negotiated their untenable living situation, I felt like we three were acting out a scene from a David Lynch script.

If I didn't have the prospect of seeing Patrick motivating me, pulling me through the hours and appointments, I am not sure how I'd have gotten through those wretched days of torrential rain and self-doubt. He in no way considered my job depressing but instead was fascinated that I got so wet in a day that my hands were pruney, laughing sincerely at stories about inconvenient poop and enraged three-legged dogs. He had grown up with pets of his own but had never before considered the depth of detail and devotion that went into caring for other people's. He listened with rapt attention when I related the events of my day, and, like my sister, made it possible for me to laugh at those things I might otherwise be inclined to cry over.

His days were so different from mine, moving as he did between brightly lit and playfully decorated offices and conference rooms, climate-controlled buildings, and cafés and microkitchens overflowing with unlimited free food and drinks. He had that desk job in front of a computer that I had so dreaded when I started down the professional pet-care path. Now that didn't sound half bad to me.

As genuine as his interest in my work was, he didn't view it as a means of defining me. He thought grad school was a super option and championed that pursuit as well. His positivity and belief that I could do whatever I wanted, be it bartending at a bowling alley or writing a book, was a salve for my dispirited soul, and my gratitude that I'd met him when I did knew no bounds.

He seemed to represent everything I'd been missing—things I needed and hadn't even realized. That weekend, he was taking me out to dinner in San Francisco, and not just for a burger or tacos. We were going to a rather fancy-looking pan-Asian restaurant that was a favorite of his. This was something he loved to do and that I'd rarely indulged in until then—sampling the finer food the city offered in abundance. These extravagances were delightfully tempered by our tradition of watching lowbrow TV on the couch at his apartment. We hadn't missed an episode of *The OC* since we'd met. Though he was bilingual and incredibly well-traveled and well-read, I was gratified that he'd been just as happy to see *Curious George* in the theaters with me over the limited-run foreign film that was showing nearby.

As I catalogued his seemingly endless interesting qualities and surprising attributes, I never took for granted that he also wasn't possessed of four paws, fur, and an inability to communicate beyond barks, wags, and licks. In him I saw a glimmer of hope—a long-absent ray of sunshine and much-needed human companionship—beyond those dismal, underemployed days of rain and uncertainty.

To: "Membership LISTSERV"
Subject: Missing Dog—UPDATE

Dear Colleagues,

Update: Missing dog FOUND! Needs immediate re-homing or foster care. Tickles, a cat- and kid-friendly Tibetan terrier, needs a forever home ASAP. Long story. Please contact me if you or someone you know can take her.

Many thanks,
Lindsey Grant, Secretary

CHAPTER TWELVE

Lost and Found

....

I had a five-day-a-week walk that was a quick commute by freeway, the client paid promptly, and they rarely if ever canceled. Sure, their Christmas tip had been Starbucks peppermint hot chocolate, but their patronage was critical to keeping a roof over my head and the lights turned on.

They lived in the Piedmont hills in a beautiful two-story Craftsman atop a long flight of flagstone steps. Frank and Diane were the prototypical rail-thin man and fleshier woman. Her struggle with weight was allegedly due to a terrible car accident that left her with irreparable back damage, a cane, and an array of painkillers. Frank had a medicated glaze to him that was probably due to some prescription of his own. They had a daughter, Mia, for whom the dog was likely meant to be a companion. I suspected it was she who had been allowed to name the dog.

Tickles was a darling Tibetan terrier. She had inquisitive eyes, was quick to reward human company with a wag and a lick, and warmed to me immediately. With the wife's back problems, regular dog walks (and apparently even getting up and down the stairs to the house) was becoming an impossibility, and the husband worked a full-time job outside of the house. Occasionally, he'd be home for one reason or another when I arrived, happening across him in the foyer or coming down the stairs looking rheumy-eyed and unsteady. I tried to be as brief and noncommittal with my greetings as possible and just get in and out and on my way with Tickles. His vacant stare gave me goose bumps.

Tickles had a dog run in the back of the house, kind of a glorified gutter where she could eat, drink, piss, shit, and work on the peanut butter kong that was thrown to her every morning after the humans had their breakfast. It was no wonder that our walks were not only the high point of her days, but that on the street she was barely manageable. She was tiny but still arguably a menace to any other canine (or cat or squirrel-like creature) that crossed her path. It was my job to exercise her to the point of exhaustion while training her to behave on-leash and learn how to control her aggression in the presence of other animals. I found that, in the six months or so that I walked her, Monday through Friday, between the hours of eleven and two, I failed on all counts.

I tried to teach her how to walk at a heel and not pull, and, when she caught a whiff of dog, to focus on me and maintain a sit, without flying into a whirling, snarling, slavering rage. It was somewhat comical to observe dogs three or eleven times her size reacting to her fits. Most of them were well mannered and responded with the canine equivalent of raised eyebrows and a discreet crossing-the-street avoidance. The owners often regarded me with some mixture of pity, distaste, or annoyance as though I had prompted her to act like such a maniac. "She's not mine," I wanted to call to them as they sauntered smugly away with their impeccably behaved companions. Cute as she was, her on-leash antics were downright embarrassing.

In spite of her dreadful behavior on our walks, I really loved Tickles one-on-one, and she was one of those rare certainties in my job. Monday through Friday, steady money, a regular check at the end of every month. This could not be more appreciated. I needed that little dog. And, it seemed, she needed me too. The little girl wasn't old enough to take her out alone, and it didn't look like Tickles was going to get much attention or exercise on the parents' time.

In concert with Tickles's regular visits, I'd been granted a brief reprieve from penny-pinching thanks to an extended pet-sitting gig for a cranky and extremely aged dachshund. It was a tricky job, requiring twice-daily visits at the very tippy top of one of Berkeley's many hills, reachable only by a single twisty road. So remote was this house that it was the penultimate mailbox from the end of the road, where the concrete ended in some overgrowth and forced the driver to make a three-point turn and return the way he or she came.

This was the very first job that I invited Patrick along for, an opportunity for him to experience my work in person. I was so excited to drive up into the hills with him, right along the Oakland/ Berkeley border, and take in the view from the top together. Up, up we went, on the curving roads fraught with hairpin turns and no guardrails, no street lamps, and barely enough room for cars going in the opposite direction to squeeze by. The houses we passed were mostly the type with the unassuming garage up top at street level, belying the sprawling houses set into the hillside below.

The view of the bay and the city of San Francisco from the client's front yard was unobstructed, the bridge looking like something from a child's erector set, and all the houses in between were little spots of life and light on a canvas. It looked unreal from so high up and left me feeling exhilarated by the height and scope of the perspective. This euphoria was quickly extinguished, though, when I turned my attentions to the yard and house itself, and the dog lurking within.

The front yard was a mystery to me, with wild, unrestrained plants, some dying and some overgrown into impenetrable thickets, accented by a rusting lawn chair or dulled, scratched garden globes. The front porch was cluttered with birdhouses, galoshes, assorted shovels, and countless cobwebby pots. It seemed to be perpetually gray and damp there. Despite the unblemished view from the edge of the property, the lot itself was heavily shaded by trees with low-hanging branches that made everything feel vaguely oppressive.

When we arrived for that first visit together, Patrick was quiet as I opened the heavy front door. He knew this client only as the reason I'd stolen away so early on recent mornings. I'd lean over him and rub his back, telling him I had to go, and he always grunted in response, claiming later that he'd never heard me and woke up thinking I'd snuck out on him.

The notion that I'd want to get away in favor of an old dachshund in an even older house made me smile inwardly. How desperately I hadn't wanted to leave his side, even if it weren't for this vicious little beast of an animal. I'd awake knowing that I was already late and she'd have made a mess somewhere in the house for me to clean up.

But this being a weekend, Patrick could come meet the reason for my hasty early morning departures. He had a softness for dachshunds, having grown up with them. There was Princess, their first, followed thereafter by Princess Two. This is beyond me, calling successive dogs by the same name. We were so gutted by the death of Biscuit that we couldn't bear to ever have another dog; we had to switch species entirely, moving on to our cat, Seal. (She was a biter, too—the only biter I ever loved.) To have a second dog, and one named Biscuit Two, would have been unthinkable.

We entered. Sure enough, I could smell that she'd done it again. She was always hiding among the clutter when I arrived, in one of a few places: under the back corner of the massive oak table in the center of the room, or beneath the lamp table that I had to reach in order to have any light. I hoped every time, as I made my way carefully toward that lamp, that I didn't step in a pile or a puddle of her mess, and I wondered for the umpteenth time why these people didn't install a light switch—or at least plug in another lamp—next to the front door.

I told Patrick to wait for me by the front door, and I got to the lamp without incident. I turned the switch, and Bitsy was revealed, snarling up at me. She was a sixteen-year-old dachshund, pure-bred, her show name something absurd like Countess Beatrice von Fluffington IX, which I heard once and dismissed straightaway. Her owner called her Bitsy for short, which worked for me.

I couldn't walk her until I located and cleaned up her mess, or messes. This was a trick, considering the living room was car-peted in a fraying, elaborately patterned rug, with flourishes and arabesques that perfectly masked a urine stain or a pile of dog shit. Luckily, this time she had soiled the hallway, which led to the down-stairs bathroom. The planks were honey wood and contrasted help-fully with the pile of poop upon them. By this time, Patrick had ventured into the living room.

"Oh my god, what a mess," he said softly, taking in the piles of books and papers that covered every surface. "Are they hoarders?"

"I have no idea. I try not to look too hard at what all is here and just focus on getting in and out without too much trouble," I said, as I leaned over with my mitt of paper towels, wrapped four-layers thick around my hand. I grabbed her little goat-pellets of poop and shoved them in a Safeway bag, spraying down the site with Trader Joe's all-purpose cleanser. I'd started to associate the pungent sage scent of the spray with this hellish job and stopped using it at my own house for its negative connotation.

With the shit all cleaned up, I started the dance of hooking Bitsy up to her leash. Holding a biscuit in front of her with one hand, I poised the leash clip with the other. She caught the scent of the treat, and, just as she was about to lunge at my biscuit hand with her sharp little teeth, I dropped it in front of her and attached the leash clip to the loop on her collar.

When I was successful, she'd fiercely gobble her treat before she realized she'd been hooked up to the leash, and I could get my hands well away from her mouth. But it didn't always go so smoothly, and she'd clamp her mean little jaws around the fingers offering the treat, or else see my other hand in her peripheral vision and snap her angular head around to chomp down on that hand instead.

I hated biters. I couldn't help but radiate anxiety when working with them, and I knew she could sense the tension coming off me. When I initially met with her owner for our consult, she'd laughed about Bitsy's biting and her boyfriend's indignation at being on the receiving end of so many painful nips. "She just prefers me," she had said airily. I realized later—too late—that Bitsy had been so well-behaved on that visit because the owner was there. But with Mommy gone, all manners and any semblance of good behavior flew right out the window. I was comforted, however, to know that I wasn't the only one she was biting. That poor, poor boyfriend.

I prompted Bitsy toward the front door and, placated by the treat, she followed. I led the way out onto the porch, down the steps, and into the snarl of a garden. I waited as long as it took for her to do her business, either in the garden or on the street. It was a certainty that if she didn't do it in my presence, she'd do it in the house before I returned for the next visit. In the evenings, the house was pitch-black inside, and I ran a greater risk of putting my foot in it. Today she waited until we were out on the street in the sodden leaves by the side of the road to squat, piss, trundle a few steps, shit, and then kick, kick, kick her stumpy legs in an effort to cover it up. Did she kick inside the house as well when she shat on the rug? I wondered. Were there pellets lurking feet from the scene of the crime that I had yet to discover, propelled by a well-placed kick from her sharp little paw?

Back in the house, I still needed to feed Bitsy, which was another dangerous endeavor. She was fiercely food reactive, and, as soon as her bowl was filled, I could not even be in the room or she would snarl menacingly, even going so far as lunging at me at with her teeth bared. I demonstrated this for Patrick's benefit, and he let out a short bark of disbelief.

"Let's just go," I said. I had screwed up and fed her before refilling her water bowl, but getting near that bowl was never going to happen now unless I waited until she was done eating. Even if I used my foot to maneuver the water bowl far enough away from her to pick it up, she would bite at my shoes until I surrendered and backed away. Crazy beast. She had enough water remaining in her bowl that I could wait to refill it when I returned.

"Ready?" I asked. He nodded, a bemused look on his face.

I went back to turn off the light.

"Why don't you just leave it on? Then you don't have to stumble around in the dark tonight," he said.

I chuckled to myself. "Client's orders," I said. "They like to save energy."

"And you honor that? At the risk of walking into furniture or stepping in dog shit?"

I shrugged. It hadn't occurred to me not to honor it. I said I would, so I did. My comfort and the convenience or practicality of the clients' requests weren't really factored; they asked, and I acted accordingly. "Well, yeah. I mean, what if they found out that I left it on?"

"Yeah, what if they did? Are they gonna fire you? From never coming here to clean up shit and get bitten by their dog?"

"Maybe!"

"And would that be so bad?" He looked back into the kitchen in Bitsy's direction and, seeming to know she was being derided, she bared her teeth right back.

I hesitated with my hand on the switch.

"No, actually that would be the best-case scenario."

So many things could go wrong with the animals or their owners' houses that were entirely out of my control; I didn't see the need to tempt fate by intentionally going against orders. Even if I reviled the mean little mongrel I was charged with watching, or I thought the clients themselves were rude or unreasonable, I had zero desire to be found in the wrong in any way at all. Or worse, to be responsible for some greater catastrophe that could have been avoided had I followed all instructions to the letter.

Many months prior, I'd taken on a cat-sitting client—not one with great long-term promise, as their regular pet sitter was unavailable and I was effectively a sub. They had seven cats, a mix of indoor only, indoor/outdoor, and exclusively outdoor. I took extensive notes on which cats were equipped with the special magnetic collar that triggered the raccoon-proof cat door, minding when they usually came inside and where their food bowls were so I could keep track of whether they'd returned and eaten throughout the week.

At my introduction, I didn't actually lay eyes on all seven cats, as some were off gallivanting in the great outdoors. But I had physical descriptions and names to go by, and the owners felt confident I'd have no problems keeping track.

But I never did see one of the cats, and its whereabouts gave me no end of anxiety. I emailed the clients about it, letting them know I was concerned, and that it didn't appear that Missing Kitty had either returned or eaten at all. Where was this mystery cat? Had she ever existed? Was this a test to see if I was paying attention?

Upon their return, they hadn't spotted her either. They were terse with me (and that's a generous word) when I gave back their keys and picked up payment, and—while I didn't expect them to— they never employed me again. I wondered what in the world could I have done differently to prevent a cat I'd never seen in my life from never showing up or coming home?

So against all logic and reason, I still turned off the light in Bitsy's living room, and we exited the house silently, the only sound the bolt of the big door sliding into place as I locked it. But Patrick's point wasn't lost on me. This anxiety I had over screwing up permeated every moment I spent with the animals anymore and seemed to cancel out any joy I derived from their company. Somewhere along the way, my baseline confidence that everything would be all right, so long as I tried my hardest to provide the best care possible for these often-complicated creatures, had been lost.

I didn't ever take care of Bitsy again after those aggravating two weeks, and not because I'd screwed up in any discernible way, or even because I refused the offer, but because she died a short four months later. She was, after all, extremely aged; it was bound to happen, and sooner rather than later.

I also didn't take Patrick on another job, not because he didn't ask, but because I'd glimpsed my work from his perspective for that short half hour and I didn't feel the need to expose either of us to that feeling again. At least not for a very long while.

It seemed increasingly apparent to me that I was unable to address my own fundamental needs simultaneous to those of the pets I spent my days, and so many nights, caring for. Following our visit with the biting dachshund, I feared Patrick had spotted my inability to manage the two—fulfillment for me, contentment for them—and that he saw this as a failure. Feeling deeply unsettled, and like my vulnerabilities had been laid bare, I finally asked him if I was right in my assessment.

"You're making a mountain out of a messy house and a mean dachshund," he said. "That dog was completely unloveable, and you're great." He kissed my forehead as though to reinforce his statement.

His innocuous comment regarding the lights in Bitsy's house—such a simple suggestion to make my life easier—had been just that. But it had prompted me to question whether I could find a way to compromise that didn't compromise the integrity of my work; to ask which was ultimately more important, my happiness or the dogs'.

More than ever, I longed for the changes that grad school might afford me: the chance to work toward my own betterment, to address some of my own needs over those of the pets I was tending to. In very concrete terms, going back to school included the entice-ment of a student health insurance plan, which meant getting back on antidepressants. The prospect of scholarships to help with cost of living, and the greatly reduced laundry and sunscreen require-ments didn't hurt, either. This would be good for mind, body, and wallet. Even as my desire to continue my studies intensified, I was starting to give up on the likelihood that any grad school was going to admit me. I had given my colleagues ample warning that this was a possibility, and one woman I contracted for was even interested in buying my client list. As the weeks wore on, though, my chances of admittance were dwindling.

Saint Mary's had been in touch, and—thanks in part, I'm sure, to my snoozing through the class visit—I'd been demoted from the wait list to the non-acceptance list. I can't say I was surprised; I'd suspected that my inexcusable in-class napping would be the killing blow to any future between me and Saint Mary's, and I wouldn't be forgiving myself for it any time soon.

I was bracing myself for a big decision ahead: either restructure my business model to compete for group-walk clients, or find a new job that didn't involve me getting pooped on or bitten or sleeping with dogs instead of this handsome man I was trying not to scare off.

To further complicate this imminent reckoning, I received a panicked email that Tickles had disappeared. Her gate was open, and she was missing. The immediate theory put forth by Diane was that some hooligan had opened the gate in one of those random malicious acts without motivation. (A one-armed man did it!) But did this hooligan steal her, too, or just release her? I wondered. She was, after all, a purebred dog. Dognapping didn't seem to be part of the leading conspiracy theory, though. Diane was convinced that between canvassing the neighborhood and searching area shelters, we'd find her. "We," meaning, of course, her and me, together.

I sent out an urgent message to the email list of fellow dog walkers, attaching a photo of the dog and a description of the situation. A few extremely kind (or bored) members wrote back and said they'd help post fliers. So on a Wednesday afternoon, during the hour that I would normally be walking with Tickles, I was tracing our various routes posting flyers bearing her image, head cocked to the side, eyes seeming to say, "Find me! Bring me home!" in a most heart-wrenching manner. I was sure we'd find her. It was only a matter of time and a question of determination, and I had both. If I lost her as a regular walk, my business and bank balance were both all but doomed.

The weekend came, with no word. No calls, no sightings, no nothing. I was not only baffled but was also starting to face the reality that that $100 a week, $400 bucks monthly, had disappeared right along with Tickles's furry little heft. I was in a bit of a panic.

Diane informed me that she had told Mia that Tickles was at the vet. She didn't tell her she was missing. Apparently she felt this worry was too much for an eight-year-old to handle, or perhaps she was avoiding an unnecessary upset if the dog could be found and the incident forgotten. I worried for this little girl, with her physically incapacitated mother, her mentally incapacitated father, and now her friendly little companion missing. She didn't even know yet! I was tempted to go over to the house, pull the little girl aside, and give her a quick but gentle life lesson on love and loss.

Then Tickles was found. She had been scooped up by a responsible citizen and taken to the Berkeley Humane Society. Diane called me in tears, overjoyed that the dog had been found and was unharmed. I, too, was overjoyed and breathed a huge sigh of relief. The terrier was safe, and so was that income.

That was on a Saturday. On Sunday, Diane called again, once more in tears. When she brought Tickles home, "back from the vet," to Mia's understandable delight, Frank declared that the dog could not stay.

He then confessed, out of range of his daughter's hearing, I hope, that he was the one who had let Tickles out. Instead of having explained calmly and rationally that the dog was too much of a burden and they would have to find another home for her, he'd crept out in the middle of the night and unlatched her gate, hoping that the rest would take care of itself.

His poor wife. She tried to collect herself on the other end of the line, but she couldn't seem to stop repeating the facts, mostly for her own benefit. She couldn't figure out how she was going to explain this to her daughter. Or forgive her husband for his callousness and cowardice, his deceit, and his shocking cruelty to both family and animal. I was with her on all counts.

For my part, I was beyond fury. I felt exhausted to my very core and could hardly fathom that this was the explanation for the past week's suspense and community effort to locate Tickles and bring her home. How, I wondered, could anyone ever safeguard against this kind of insane scenario? Baffled as I was by this unexpected turn of events, I redoubled my efforts, now to find Tickles a new home, instead of just to find her at all. With some bribery in the form of free walks and some convincing that she was the best and cutest and most loveable and perfectly sized dog, I was able to work out a temporary fix with a friend and nearby neighbor who had been wanting a dog.

So now, not only was I out many hundreds of dollars, I was also out a walk slot that would still go to Tickles, with no monetary benefit to me.

This didn't last for long. I kept up my end of the bargain, walking her every day, but Tickles was not a fit for this home. My friend and her fiancé didn't take too well to Tickles's barking; her assertive terrier nature, and her manic aggression toward all other fauna; or her general need for feeding, cleaning, love, and attention. They hadn't actually wanted a dog, but maybe a stuffed animal or wind-up toy.

This time, my friend led the charge in finding Tickles a more-permanent home; she was that motivated to offload this pitiably displaced dog. A few interested parties came and met Tickles, and my friend ultimately sent her home with a woman who lived on a farm in Orinda. On the day she came to pick up Tickles, she brought her daughter along, as well as a bouquet of flowers as thanks for this precious, purebred, and entirely free pet she was taking home. She was also in tears, so happy was she to have found a dog just like their little Maisey, may she rest in peace. I hoped with all my heart that Tickles would be happy on the farm, and that she would fill the terrier-shaped void in these people's lives.

I still wonder what in the world Mia's parents ever told her, and how many years down the road she would learn the bizarre truth about her father's betrayal. That family had seemed on shaky ground enough with the dog, but I had a hard time imagining how they'd be able to move beyond the unfortunate series of events leading to the permanent loss of their pet.

Ian's and my lease on our shared apartment was ending, and he'd declared that he was moving to San Francisco. Not knowing where I might move and with whom, if anyone, and what job, if any, might pay for this next rental, I started packing up our apartment.

Though I liked to daydream about my future with Patrick, I knew that it did not realistically include moving in together within the next month or so. He already lived with three other guys and had recently declared that it felt too soon for me to meet his parents. They had been in town and had taken him out to a dinner at which his sister and brother-in-law were present, while I was not. I appreciated his forthrightness, and even kind of believed him when he explained that it was less that he wasn't ready for them to meet me and more that he wasn't prepared for me to meet them. He wanted to keep me to himself for just a bit longer. Either way, whether this was the truth or not, moving into a love nest with him was off the table for the present.

I was in the bathroom preparing for a day like any other when I missed a call. I'd heard the ring from my perch upon the toilet and figured it was either a client canceling or positing a special request, or else my parents checking in on how things were going. In either case, I could call them back.

When I checked my phone, however, the missed call came from a new number, one not saved in my contacts. I opened my voice mail.

Perhaps it was the unfamiliar voice or sheer disbelief that was impeding my brain's ability to translate the words I was hearing into any semblance of meaning, but I had to listen to the message a full three times before flipping my phone shut and letting out a strangled squeal.

According to the professor who had phoned, I'd been enthusiastically admitted to the graduate creative writing program at Mills College, right down the highway in Oakland, and she was extremely excited to speak to me about it. But not, I'm certain, as excited as I was.

Things happened very quickly after that phone call. I, of course, called to accept and, thereafter, began applying for loans and financial aid opportunities. I started looking at apartments near campus and investigating part-time jobs that might help subsidize living costs and whatever tuition wasn't covered by student loans.

I decided, easily and definitively, that despite what my clients and colleagues said, I could not—would not—continue walking dogs simultaneous to getting my degree. Of course, as soon as I started sharing the news of my acceptance with my immediate dog-walking community, offers started pouring in. To run the business side of a colleague's operation. To continue on with a few select clients at better pay. To take a part-time job at a newly opened doggie day care.

I had to laugh at the irony. Where was this bounty a mere month ago? Nevertheless, I felt convinced that my number one priority had to be school and my studies. That door had opened, and I needed to close this one firmly and for good.

My mom came out to help me apartment hunt, and we signed the lease on an idyllic little studio a mile from the school. It was an in-law unit that backed up to an organic garden maintained by my new landlords. The deposit in hand, they said I could move in just as soon as I was ready.

My professional sendoff was unexpectedly sentimental. As focused as I'd been on extricating myself from the mess of my chosen profession, I'd overlooked the fondness my clients might feel for me, their animal nanny, fairy godmother to all their pets' many and obscure needs.

The greyhounds' owners gave me pillowcases screen-printed with pictures of Flannel's and Salvador's disembodied heads. The shelties' mom gave me a bottle of Beano and Discus–branded wine, the label a photo of the dogs looking wistful, their fur windblown. On my last day of walking Foxy, the owners left a note with a box of chocolate, reminding me that their offer of continued employment stood should I ever change my mind. The grandest gesture, by far and away, was the Montblanc pen engraved with my name that my colleagues at the pet-sitters' association chipped in to buy me in thanks for my services as secretary.

Random as many of these gifts were, they were such fitting tokens of my time as an animal nanny. Each represented in its own weird way the significance of earning my colleagues' and clients'—and their beloved pets'—trust, symbolizing the rare privilege of being allowed to participate in the terrifically unique and always deeply personal relationship between parent and fur baby. However outlandish or uncomfortable the details had gotten, it had been an honor.

Still, after two years of making my own issues and priorities secondary to those of the dogs and birds and cats and their owners that I'd catered to, I was looking forward to an entirely different life landscape. The challenges I might face in grad school, living on my own, and pursuing this relationship with Patrick would likely feature far less fur and fewer feathers, not as much getting locked out of places, and (hopefully) not as many creative approaches to shit collection. Whatever the demands, however devious and unexpected the obstacles, I was ready. I had, after all, done my own self-employment taxes. Twice. Surely that alone prepared me for the worst life might throw my way.

On my final day of dog walking, after I'd returned all the clients' keys and hung up the last leash, I uncorked the bottle of Beano and Discus wine. We raised our glasses, Patrick and I, to what I was leaving behind, and all that lay ahead.

A fellow dog walker I'd contracted for bought my client list after all, employing some formula for walks per week over rate I charged by length of patronage or some such method to determine the price. It was a humble amount, but that sum, coupled with Ian's and my recently returned and divided security deposit, was enough for a plane ticket.

Patrick had had a change of heart, and, instead of inviting me to a family dinner, he'd asked me to join him and roughly thirty of his relatives for a family reunion in Tuscany. A better way to celebrate the end of one era and the beginning of another, and to spend the spoils of two years' work, I couldn't imagine.

So I said yes.

Acknowledgements

....

I t sure takes a village to get down this path to publication, and there are scores of people responsible for getting me here. Ms. Frost, Mrs. Pugh, Professor Williams, Dr. Craig, and Ms. Fink, you maintained belief in my ability to churn out words that other people might want to read, even when I myself had no faith. To my mentors Pam and Jerry, thank you for your tutelage, and for trusting me to walk alongside you. I surely didn't deserve your investment in me, but I'll appreciate it forever. C, B, B & AJ: You truly adopted me, minus the legalese. You're my second family and I love you always. My professors and fellow writers at Mills saw the earliest attempts, and I am forever indebted to them for the direction they offered. To the inimitable Chris Baty, you are the reason I was ever published in the first place. Thank you for your encouragement, and for introducing me to the best agent I could ever hope for. Lindsay Edgecombe, your patience and prescient guidance has been invaluable to me. To my writing partner of many (many) years, Elizabeth Gregg, I am so grateful we kept dragging each other out into the bars and cafes of

Oakland to hack away at our slowly evolving projects. To my editor at Seal, Laura Mazer, you're a dream to work with. The pleasure of publishing with you and your colleagues at Seal has been entirely mine. My family at National Novel Writing Month, colleagues and participants alike, have provided incomparable inspiration. You are my heroes—every one of the many hundred thousands of you. To the best Jen and Mark, Nora, and my cousin John, you were the first and only readers of the original final draft so many iterations ago, and you never once laughed in my face. I can't thank you enough for your time and kindness and praise. Sister mine, you have been championing my words from the time I was writing wallpaper books and publishing poems in the Fernbank newsletter. Thank you for being my biggest fan, since before I understood what that meant and how invaluable it would ever be to me. I am yours right back. And as the dedication states, I would be lost if not for the endless love and gentle prompting from my exquisite parents. I only wish Dad were here to hold this book in his hands. He was passionate about seeing it published and he very narrowly missed getting to see the finished product. I intentionally saved this book's MVP for last. Pat, my love, my BFF and prince husband, you have supported me at every twisty turn along the way, and this book simply wouldn't have made its way into the world without you. Thank you for believing in me and doing everything in your power to make my dreams come true. It's your turn now.

About the Author

....

Lindsey Grant is the former program director for National Novel Writing Month (NaNoWriMo.org), a nonprofit organization that encourages writers of all ages and backgrounds to pen the draft of a novel during the month of November. She co-authored the writer's workbook *Ready, Set, Novel!* (Chronicle Books) and holds an MFA in creative nonfiction and English from Mills College in Oakland, California. She lives in Zurich, Switzerland, with her husband and their cats, where she writes, tries to speak German, and blogs about her attempts to assimilate.

Selected Titles from Seal Press

Cat Women: Female Writers on Their Feline Friends, by Megan McMorris. $14.95, 978-1-58005-203-0. From a tale about how rescuing a stray cat ended up saving a friendship to an unapologetic piece by a confirmed—and proud!—crazy cat lady, this collection of essays ranges from thought-provoking and heartrending to laugh-out-loud funny, delving into the many ways these often aloof little divas touch our lives.

Gawky: Tales of an Extra Long Awkward Phase, by Margot Leitman. $16.00, 978-1-58005-478-2. Tall girl Margot Leitman's memoir is a hilarious celebration of growing up gangly, a cathartic release of everything awkward girls are forced to endure, and a tribute to a youth that was larger than life.

Woman's Best Friend: Women Writers on the Dogs in Their Lives, edited by Megan McMorris. $14.95, 978-1-58005-163-7. An offbeat and poignant collection about those four-legged friends girls can't do without.

The Quarter-Acre Farm: How I Kept the Patio, Lost the Lawn, and Fed My Family for a Year, by Spring Warren. $16.95, 978-1-58005-340-2. Spring Warren's warm, witty, beautifully-illustrated account of deciding—despite all resistance—to get her hands dirty, create a garden in her suburban yard, and grow 75 percent of all the food her family consumed for one year.

Why We Ride: Women Writers on the Horses in Their Lives, edited by Verna Dreisbach, foreword by Jane Smiley. $16.95, 978-1-58005-266-5. Verna Dreisbach collects the stories of women who ride, sharing their personal emotions and accounts of the most important animals in their lives—horses.

Follow My Lead: What Training My Dogs Taught Me about Life, Love, and Happiness, by Carol Quinn. $17.00, 978-1-58005-370-9. Unhappy with her failing love affair, her stagnant career, and even herself, Carol Quinn enrolls her two Rhodesian ridgebacks into dog agility training—and becomes the "alpha dog" of her own life in the process.

Find Seal Press Online
www.SealPress.com
www.Facebook.com/SealPress
Twitter: @SealPress